THE EXPENDABLE FUTU

M000118155

U.S. Politics and the Protection of Biological Diversity

Richard J. Tobin

Duke University Press   *Durham and London 1990*

Second printing, 1993
© 1990 Duke University Press
All rights reserved
Printed in the United States of America
on acid-free paper ∞
Library of Congress Cataloging-in-Publication Data
appear on the last page of this book.

To Noah

Contents

Abbreviations

APHIS	Animal and Plant Health Investigation Service
BCF	Bureau of Commercial Fisheries
BLM	Bureau of Land Management
BOR	Bureau of Reclamation
BSFW	Bureau of Sport Fisheries and Wildlife
CEQ	Council on Environmental Quality
CITES	Convention on International Trade in Endangered Species of Wild Fauna and Flora
CREWS	Committee on Rare and Endangered Wildlife Species
DLNR	Department of Land and Natural Resources (Hawaii)
DOE	Determination of Effects
EPA	Environmental Protection Agency
ESTB	Endangered Species Technical Bulletin
FDAA	Federal Disaster Assistance Administration
FR	Federal Register
FWS	Fish and Wildlife Service
GAO	General Accounting Office
LWCF	Land and Water Conservation Fund
MMC	Marine Mammal Commission
NEPA	National Environmental Policy Act
NMFS	National Marine Fisheries Service
NPS	National Park Service
OES	Office of Endangered Species
OMB	Office of Management and Budget
OTA	Office of Technology Assessment

PLLRC   Public Land Law Review Commission
TVA   Tennessee Valley Authority
USFS   United States Forest Service

Preface

Evaluating any public policy is difficult. The task becomes even more so when the goals of the policy evolve, when several agencies share responsibility for implementation, and when these agencies must cope with inadequate organizational resources, uncertain public support and, too frequently, political antagonism and interference. Such is the case with efforts to evaluate the accomplishments of the Departments of the Interior and Commerce in their quest to implement the Endangered Species Act.

The legislative ancestors of this act initiated efforts to prevent premature extinction in 1966, but much has changed since then. Assessing the likely need at that time, officials responsible for these efforts probably envisaged a relatively small program. Protection of species threatened with extinction was neither a pressing scientific nor a hot political issue. By the early 1990s, however, the pace of human-caused extinctions had accelerated to rates not previously known in the United States or in most of the world. Scientists familiar with the problem share a growing fear that these extinctions may lead to an irreversible environmental catastrophe.

Biological and ecological scientists are focusing increased attention on the causes and consequences of the premature loss of biological diversity, but most social scientists are consistent in their total neglect of the topic. This book attempts to redress this neglect, at least from the perspective of policy analysis and political science. As the book suggests, the protection of biological diversity raises fascinating economic, political, and institutional issues that will not soon fade.

For most authors books take too long to research and write, and the people due an acknowledgment begin to get impatient. In the present case they probably have some justification; perhaps they will find their reward in the pages that follow. Among those to be thanked for their assistance and encouragement are Ed Herman, Dave Johnson, Peter Gold, and Sheila Conant. The U.S. Fish and Wildlife Service (FWS) is thanked and commended for its complete openness and continued responsiveness to myriad requests for information over several years. At times the book is critical of the agency, but no one can question the integrity or good intentions of the people involved with the Endangered Species Act. Service employees who were particularly helpful include John Fay, Larry Thomas, Nancy Sweeney, Vicki Finn, Clyde Schnack, and Bob Graves. John Fay carefully read an earlier draft of the book and provided especially valuable insights from his perspective as a botanist with the FWS. Other agencies that were helpful during the research process include the Marine Mammal Commission, the Army Corps of Engineers, the National Marine Fisheries Service, Hawaii's Department of Land and Natural Resources, and California's Department of Fish and Game.

Several people offered useful criticisms after their reading of the draft manuscript. These include Howard Wolf, Department of English, State University of New York at Buffalo; Michael J. Bean, Environmental Defense Fund; Michael R. Sherwood, Sierra Club Legal Defense Fund; and Roger L. Di Silvestro, National Audubon Society. An abridged and earlier version of chapter 9 appeared as "Interorganizational Implementation of the Endangered Species Act: A Hawaiian Case Study," in *Journal of Land Use and Environmental Law* 4 (Winter 1989): 309–35. In addition, I would like to thank the anonymous reviewers for Duke University Press, whose comments were particularly helpful.

Finally, and most important, my wife Dori deserves eternal accolades for her patience, support, and encouragement. I sincerely hope I will continue to be the beneficiary of her virtues once research begins on the next book.

Richard J. Tobin

1 Biological Diversity as an Ecological Issue

Working in the desert of California's Death Valley in the late 1940s, a biologist noticed some small fishes in two brine-filled hot springs. The fishes were different from others previously found, so the biologist described them scientifically and subsequently reported the discovery of a new subspecies —the Tecopa pupfish. The subspecies had probably existed for thousands or perhaps even millions of years, but neither its discovery nor the scientist's description of the pupfishes was unique. Millions of species and subspecies of plants and animals exist, but most are unknown and unnamed.[1] The consequence is that most species remain to be discovered, collected, and described scientifically. Unlike the discovery of a new planet, the discovery of a new species is common.

Equally common are efforts to transform the environment to accommodate human life-styles. These efforts were fatal for the pupfish. Some adventurous entrepreneurs decided that the hot springs would provide an ideal location for several bathhouses. Without considering the needs of the pupfishes, the two hot springs were combined and their flow was altered during construction in the mid-1960s. Within a few years both the U.S. Department of the Interior and California's Department of Fish and Game announced that the alteration of the pupfishes' sole habitat jeopardized their continued existence. By 1972 none of the fishes could be found. Extensive searches in later years also failed to locate even a single pupfish, and, in 1982, it became the first animal to be removed from the Interior Department's list of endangered species because of extinction.[2]

Modest changes during construction of the bathhouses would have saved the pupfishes; their extinction had been "totally avoidable."[3] Regardless of whether extinction could have been avoided, the demise of the pupfish went largely unnoticed. Radio and television networks neglected the extinction; no flags were lowered to half staff; no protesters complained to the president about an ecological catastrophe; Congress continued its ponderous deliberations without hesitating for a moment of silent prayer.

The extinction concerned few people, perhaps understandably so. Never had the pupfish served a practical interest or bettered anyone's life. The Tecopa pupfishes might be slightly different from their aquatic colleagues, but they were useless to humans, or so most people would conclude had they even known of the fishes' obscure and isolated existence. Further justification for this view came when the pupfishes' extinction apparently caused no changes in life-styles or environmental quality. Indeed, local residents were probably happier without the pupfish. At least their town of Tecopa had several bathhouses that could profitably use the two previously undeveloped hot springs that once provided the pupfishes' only habitat.

Despite the apparent lack of concern about the pupfish, its extinction provides an example of a growing and much larger problem. Extinction is a fate from which no species can ever escape, but natural extinctions are infrequent. Among all species—ranging from the smallest insects and invertebrates to the largest mammals—one natural extinction occurs every year or so.[4] Many people believe that dinosaurs became extinct over a short period; using geological time scales, perhaps they did. As one scientist has estimated, however, during the "great dying" of the dinosaurs, only about one species was lost every one thousand years. Equally important, although the rate of natural extinctions fluctuates, the evolution of new species has historically exceeded the rate of extinction.[5] Thus the total number of species has risen over time. There is no reason to expect an end to this trend unless other factors intervene to interrupt the evolutionary process.

Human activities provide just such a factor, and one that is becoming increasingly important. Human activities led to the extinction of about seventy-five species of birds and mammals, or about one species every four years, between 1600 and 1900.[6] A like number of birds and mammals became extinct because of human interventions during the first half of this century. Since then, and particularly since the 1970s, the number of human-caused or anthropogenic extinctions has soared to the point where they overwhelmingly exceed natural extinctions. In the mid-1970s scientists estimated that about one hundred species were lost per year due to human activities.[7] Estimates of expected extinction rates for the next few decades are much higher.

As they must, estimates of extinctions due to human activity vary considerably. One biologist, Edward O. Wilson, has speculated, however, that the extinction rate in the mid-1980s was accelerating rapidly and was at least four hundred times the natural or background rate. Other specialists contend that anthropogenic extinctions could be more than one thousand times as high as the background rate by 2000.[8] Matching human-induced rates with the number of predicted extinctions is problematic and occasionally controversial, but some estimates place the number of species lost at the hands of humans at as many as twenty to fifty per day. This pace is expected to quicken considerably. By the end of this century the number of species lost per day could be in the hundreds. The sum of all species will have been reduced by as many as 500,000 to 2,000,000, perhaps even more, depending on the total number of species that exist.[9] Other scientists are reluctant to specify the number of likely anthropogenic extinctions, preferring instead to forecast the percentage of all species that will become extinct because of humankind. The percentages range from about 10 to 25 percent for the short term and as high as 50 percent over the next fifty to one hundred years.[10] So dire is the anticipated situation that some scientists believe that all mammals, except humans and domesticated species, could become extinct within the next few decades.[11] This would mean the end of whales, wolves, leopards, giraffes, and grizzly bears, to name only a few of many.

Estimates of extinction rates become more meaningful when one considers the source of these forecasts. Environmentalists are often accused of overstating their case. To these environmental jeremiads, only an exaggerated list of ecological nightmares catches policymakers' attention and causes them to act. Predictions about extinction rates are often debated, but those making such estimates are often experts in their scientific disciplines. Equally important, the estimates are sufficiently compelling that national and international agencies are beginning to recognize the predicted rates as an area of major concern.

In 1980, as an example, the U.S. Council on Environmental Quality (CEQ) published *Global 2000*, a projected assessment of environmental quality at the beginning of the twenty-first century. The report calculated that as many as 20 percent of all species could vanish within two decades.[12] Although the U.S. Office of Technology Assessment (OTA) was reluctant to specify projected extinction rates in its 1987 report on the loss of biological diversity, the report emphasized that the problem is one of "crisis proportions."[13] In a report to the United Nations General Assembly in 1987, the World Commission on Environment and Development similarly concluded that "species are disappearing at rates never before witnessed on the planet."[14] Human-caused extinctions, the commission emphasized, are hundreds of times higher than background rates but "could easily be thousands of times higher."

Of the species being lost because of human activities, most are found in the tropical forests of Africa, Latin America, and South and Southeast Asia. These forests contain most of the world's biological diversity as well as most of the species that remain scientifically undiscovered. It is in the tropics that the crisis of extinction is most pronounced.

The situation in the United States is not as dire as in some other parts of the world, but human-generated extinctions in the United States are no less a concern than elsewhere. Between 1880 and the early 1940s, at least forty native species of birds, fishes, and mammals disappeared.[15] By the 1970s about one hundred domestic species of fish and wildlife were considered to be in danger of extinction, and about two species were thought to be lost to extinction each year.[16] Increased attention to the extinctions in the 1970s revealed that these estimates were far too conservative.

When the U.S. government published its first formal list of native endangered species in the late 1960s, about one hundred species were on it. By 1990 an updated list included nearly six hundred domestic species. Had the process of listing species been simpler and less costly, the list could easily have been five to ten times longer. The Nature Conservancy estimates the number of species on the verge of extinction in the United States to exceed two thousand.[17] Native plants are particularly vulnerable. Of the approximately 25,000 U.S. plant species, over 10 percent are believed to be close to extinction in their wild habitat. Other native species have been less fortunate—they have become extinct even before being discovered.

In some states, notably Hawaii and California, the problem of extinction-prone species is especially severe. The number of plants in California considered to be in danger of extinction in 1985 exceeded six hundred. Nearly all of Hawaii's native flora is in jeopardy,

and hundreds of the state's endemic species are candidates for classification as endangered. With the exception of the Mascarene Islands
in the Indian Ocean, no other region in the world has lost as many
native species of birds as has Hawaii.[18] Of its remaining endemic bird
species, the continued existence of three-quarters is in doubt. Some
have not been sighted in years.

Although the situation in the United States is not as grim as in
some countries, what happens there affects the protection of biological diversity throughout the world. For this reason the present study
assesses the protection of this diversity in the United States. Considerable justification exists for this choice.

As one of the world's wealthiest countries, the United States can
afford, better than most countries, the costs of protecting and preserving its plants and animals. Moreover, how the United States responds
to the plight of its own species serves as a barometer of American
attitudes toward and policies for the protection of such species elsewhere. If American policymakers display limited concern for native
flora and fauna that are in jeopardy of extinction, then it is unreasonable to expect that greater attention will be devoted to foreign
species. Only as concern for domestic species increases will concern
also increase for foreign species on the brink of extinction.

This is an important consideration. The United States invests millions of dollars annually in development projects in many Third
World countries. Concern for the environmental consequences of
these projects often hinges on the values of the agency sponsoring
the projects. Only in the 1980s did the U.S. Agency for International Development begin to mitigate the environmental damages of
its overseas activities. The agency now has a reasonably good record
of protecting environmental values, but this concern is not universally
shared among its sister institutions. The CEQ has identified several
public development institutions that "remain largely indifferent to
the environmental effects of their programs. . . ."[19] Furthermore, according to CEQ, the Asian, African, and Inter-American Development
Banks have poor environmental records, at least partly because the
U.S. government has not exercised its considerable influence.

Americans' vast appetite for such imported natural resources as
timber, metals, minerals, and petroleum as well as food also means
that habitats of species found outside the United States are altered
or developed to benefit American consumers. Using timber as an
example, Norman Myers calculated that U.S. consumption of tropi-

cal hardwoods increased ninefold between 1950 and 1973.[20] Most of these hardwoods came from forests in Southeast Asia, and probably would not have been chopped down except to satisfy compelling demands for such necessities as wooden floors, snack trays, and salad bowls.

The United States likewise finds itself dependent on other countries to provide vital raw materials. Americans imported more than 90 percent of their supplies of cobalt, bauxite, platinum, tantalum, manganese, strontium, columbium, and sheet mica throughout the 1980s. Imports accounted for more than 50 percent of another dozen metals and minerals, including tin, zinc, nickel, and chromium. These materials undoubtedly improve the quality of American life and make possible items as mundane as electric fuses and light bulb filaments and as arcane as high-temperature alloys and parts for nuclear reactors. At the same time, however, mining these metals and minerals can have devastating effects in areas that are rich in biological diversity. Malaysia, one of the world's largest producers of tin and its largest single exporter to the United States, had over five hundred square miles of tailings from tin mines in the early 1970s.[21] To produce the four million short tons of zinc imported into the United States between 1983 and 1987, more than 48 *billion* tons of rock had to be mined, processed, and disposed of.[22] Other examples of Americans' dependence on imported resources exist, but the situation with timber, metals, and minerals reveals that the United States is a major cause of environmental change outside of its borders.

Still another reason for emphasizing the United States involves its role as one of the world's leaders in environmental conservation. In many areas, such as the use of environmental impact statements, the United States is a model for much of the world. Such is also the case with the protection of endangered species. The United States, along with Japan, Russia, and Great Britain, signed the first international agreement to protect wildlife in 1911. It was the United States that convened an international meeting to discuss the regulation of international trade of fish and wildlife species that are in danger of becoming extinct. The result—the Convention on International Trade in Endangered Species of Wild Fauna and Flora (CITES)—was developed in 1973. The United States was the first of nearly one hundred nations that have ratified the convention. Here again, what happens in the United States affects the success of efforts to regulate the international trade of endangered species. Despite its good intentions, the United States offers the world's largest market for wildlife. Illegal imports of live endangered species into the United States are a booming business involving more than $100 million per year.[23] When one adds the millions of skins, shells, horns, furs, and feathers from deceased animals, the value of illegal imports rises still further.

In sum, although most species, including the endangered ones, are located in the tropics, events in the United States often determine the fate of species worldwide. This occurs either because of Americans' patterns of consumption or because U.S. environmental policies and their implementation serve as a standard against which other nations judge themselves. Consequently, an evaluation of public policies for the protection of native American species can serve as an indicator of the likelihood that human-caused extinctions will be halted, reduced in number, or significantly increased, not only in the United States, but elsewhere as well.

Causes of Endangerment and Extinction

When asked, most people would probably say that overhunting is a major cause of endangerment and extinction. In the nineteenth century the American bison almost became extinct because of over-hunting. Although the last passenger pigeon died in a zoo, the species was a victim of excessive hunting. Today, other species find their continued existence in doubt because of hunting, but these tend to be large game species or animals that are valued for their alleged medicinal powers—like the rhinoceros—or because of some other desirable attribute—ivory from elephants or the novelty of owning tusks or penis bones from American walruses.

In spite of the conventional wisdom, hunting provides an explanation for the plight of relatively few endangered species. Most endangered species are victims of unintentional harms rather than of direct and conscious actions, such as those associated with hunting. Some plant species, for example, are naturally rare and live only in a single, isolated area. One shortsighted episode, like the careless use of an off-road vehicle, can mean the demise of such a plant species. Hawaii's flightless birds became extinct because of the introduction of exotic predators that destroyed eggs and nests. Exposure to pesticides and other chemicals threaten still other species.

Among all causes of endangerment and extinction, however, the leading cause is the destruction or modification of species' habitats. Most of this alteration is due to entirely legitimate developmental activities, such as farming, or the construction of houses, highways, and shopping malls. The activity is well intended, and harm to the habitats of potentially vulnerable species is usually an unexpected and unintended consequence of seemingly reasonable demands for development.

Growing human populations and expectations for increased economic activity, some would argue, are natural precursors of development. Given an incentive to develop and to enhance one's wealth or well-being, land-use patterns are typically dictated by how much income and profit can be generated rather than land's ecological value in an undeveloped state.[24] Developers thus find no need to trouble themselves with seemingly extraneous concerns, like biological diversity, unless required to do so. From the perspective of vulnerable species, this is an unfortunate result. For many species, especially plants and animals that are endemic to certain areas, their habitat is a highly localized one. There may be no other place on earth where the species is found or where it could survive if transplanted. Its habitat is unique in the sense that it or, more specifically, the ecological community of which the species is a part, offers exactly what the species needs to exist. The community provides a soil of a certain alkalinity, water of a certain salinity, the only host plants upon which the species feeds, protection from predators, a special mix of temperature, rainfall, and humidity, or literally scores of other features in some unparalleled combination. When this habitat is altered or eliminated, the species may face premature extinction from its natural habitat.

The requirements of Kirtland's warbler well illustrate how closely tied a species can be to its habitat. This yellow-breasted songbird, which conveniently spends its winters in the Bahamas, nests only in northern Michigan among young jack pines that are eight- to twenty-years old in dense stands of eighty acres or more, growing on Greyling sand.[25] This habitat was created as a result of wildfires, which cleared the land for new growth and ensured the species' continued existence. Today, fires are intentionally set to accomplish this goal. (Such efforts can be extremely dangerous. The U.S. Forest Service (USFS) once started a "prescribed burn" to benefit the warblers. The fire was to be extinguished after it had burned two hundred acres, but the fire raged out of control. Before it could be stopped, 25,000 acres had been burned, forty-one homes had been destroyed or damaged, and one fireman was killed).

The Ash Meadows area, northwest of Las Vegas, Nevada, provides another striking example of an extraordinarily specialized habitat. The area probably has the highest concentration of endemic species in the continental United States. These include fishes, plants, insects, and mollusks that are found in a desert wetland in one of the world's most arid regions.[26] Among the area's several fishes, the Devil's Hole pupfish survives in a single, isolated, spring-fed, limestone pool about fifty feet below the land surface with a constant temperature of 92°F. The National Park Service (NPS) believes this to be the "most restricted environment of any animal in the world."[27] So distinct are the pupfishes that they have been called the "Darwin finches of the desert"; because of their isolated genetic development, the fishes do not resemble any other living species.

The Ash Meadows' species are both unique and endemic, but their situation is similarly repeated in thousands of other places. Disruption of fragile ecosystems jeopardizes the continued existence of many species that cannot survive in any other place. Once an ecosystem is altered or destroyed, some, many, or all of its endemic species may disappear forever. Even slight modifications can imperil a species. Some species can migrate from altered habitats, but this does not necessarily ensure their survival. Many large mammals often have large ranges and need huge areas in order to survive. Tigers illustrate the case well: as much as 16,000 square miles of undeveloped habitat are required to support a population of as few as four hundred tigers. Such an area is only slightly smaller than the combined areas of Massachusetts, Connecticut, Rhode Island, and Delaware.

In short, destruction or modification of habitats provides the best explanation for the plight of most endangered species, not only in the United States, but throughout the world. Though the causes of the destruction will differ from one region to the next, the consequences for species are likely to be the same.

Consequences of Premature Extinctions

Since the last Tecopa pupfish died in the 1970s, has anything changed? Has the species' extinction made a difference to anyone? If there is a difference, how many people know what it is? Why be concerned about the fate of the pupfish? After all, among all species that have ever existed, at least 90 percent are now extinct. All species eventually end their evolutionary journey, so why worry when human progress helps nature along its preordained path?[28]

There are many answers to this question. Though the pupfish may have appeared insignificant next to human ingenuity, Edward O. Wilson emphasizes that a species:

> . . . is not like a molecule in a cloud of molecules. It is a unique population of organisms, the terminus of a lineage that split off thousands or even millions of years ago. It has been hammered and shaped into its present form by mutations and natural selection, during which certain genetic combinations survived and reproduced differentially out of an almost inconceivably large number possible.[29]

Wilson adds that "each species of higher organism is richer in information than a Caravaggio painting, Bach fugue, or any other great work of art." Even the common mouse, a friend to few, could provide enough genetic information to fill every edition of the *Encyclopedia Britannica* published since 1768! This means that each time humans hasten an extinction, they forfeit information of inestimable value.

Wilson further emphasizes that the premature loss of even a single species can be a signal event, a judgment by humans that a species no longer deserves to exist. When humans exterminate a species, they provide a commentary about their respect for nature and the environment. People understandably find it difficult to relate to most life forms, but consider human concern and sympathy for their closest genetic relatives in the animal kingdom, chimpanzees. If one compared the genetic composition of humans and chimpanzees, the similarities would be far greater than the differences—about 99 percent of human genes are identical to the corresponding set in chimpanzees. Of the genes that are different, they can be identified only with the assistance of high-resolution photography. The slight differences between chimpanzees and humans do not make the former any less important or the latter any more desirable. Some research suggests just the opposite. In certain areas the intellectual abilities of chimpanzees exceed those of humans.[30] Despite chimpanzees' intelligence and similarity to humans, the former have been deemed expendable. Over the last thirty years humans have expropriated and destroyed most of the chimpanzees' habitat, or hunted and captured them, and they are now highly vulnerable to extinction.

Chimpanzees provide one cause of concern and one indication of human attitudes toward nature. The larger problem, however, is not the extinction of a single species, but rather the premature loss of thousands or possibly even millions of plants and animals. Extinction of a single "insignificant" species might go unnoticed, but humans risk the loss of the enormous benefits that species now provide or that they may be capable of providing at some future date.

Perhaps most important to humans, species provide free and irreplaceable ecological services.[31] No one is independent of the need for these services. Species are essential parts of ecosystems, upon which all life depends. When species are deleted from their ecosystems, the latter may be unpredictably disrupted to the detriment of people everywhere. People take for granted the benefits of ecosystems and often fail to appreciate humans' complete dependence on them. Effectively functioning ecosystems are important for the disposal of wastes, the cycling of nutrients, the control of climates, the regulation of freshwater supplies, the generation and maintenance of soils, and the quality of the atmosphere. The extinction of a single species rarely causes the complete disruption of an ecosystem, but such a loss can lead to entirely unexpected consequences. Furthermore, all species in an ecosystem are mutually dependent, so extinction of one species can mean the subsequent loss of dozens or even hundreds of dependent species.[32]

Species similarly provide benefits because they serve as unsung heroes in the control of pests and diseases. Species in natural ecosystems control all but a small percentage of the organisms capable of competing with humanity for food or of doing harm to humans by transmitting diseases.[33] Beyond the direct control of pests and diseases, species offer an unlimited reservoir of present and potential direct benefits. Ecosystems provide food and the genetic diversity that allows agricultural experts to develop new crops and disease-resistant strains of corn, rice, wheat, and potatoes. When a leaf fungus invaded American cornfields in 1970, destroying about one-fifth of the crop, nature came to the rescue. Scientists discovered a fungus-resistant strain of corn in Mexico that was crossbred with some strains already in widespread agricultural use in the United States.

Plants are also the source of many medicinal drugs now in use, and about half of all prescription drugs in the United States come from wild organisms. The commercial value of these drugs is believed to exceed $14 billion per year.[34] The Madagascar periwinkle plant is a source of vincristine and vinblastine, two alkaloids that are effective in the treatment of Hodgkin's disease. Another plant genus, *Rauwolfia*, provides reserpine, a medicine used to control high blood pressure. Two other heart-related drugs, digitalis and digoxin, which come from a species of wild snapdragons found in Europe and North Africa, have saved the lives of millions of Americans.

Wild plants are not alone in their ability to offer medicinal remedies to human ills. Certain snake venoms provide nonaddictive pain killers while blowfly larvae secrete a substance that encourages the healing of deep wounds.[35] Sponges, baboons, monkeys, butterflies, chimpanzees, sea urchins, armadillos, and desert pupfish provide a few examples of the many animals that are vital to biomedical research. The primates are particularly important because they allow the production and testing of drugs and vaccines and research on AIDS, cancer, leprosy, hepatitis, and cardiovascular diseases.

Species likewise offer research opportunities for scientists studying genetics and evolution. Wild species provide medical and scientific benefits as well as industrial and commercial ones. Certain dyes, fats, gums, oils, waxes, resins, tannins, cosmetics, and insecticides come from wildlife. Some plants provide substitutes for products that many people would prefer not to harvest. Liquid wax is used in the manufacture of products as diverse as pet foods, cosmetics, carbon paper, and electrical insulation. For most of this century, sperm whales provided the only natural source of liquid wax. In the late 1960s the United States was importing nearly fifty million tons of the whale's wax annually, much of which was used as lubricants for automobiles. The wax had so many important uses that the U.S. government stockpiled it in case of a national emergency. Sperm whales were nearly extinct by 1970, however, so it became imperative to find alternative sources of the liquid wax.

The jojoba shrub, found only in the deserts of northern Mexico and the southwestern United States, provided the solution. Jojoba seeds produce the only liquid wax known to exist in the entire plant kingdom. The wax may be a miracle substance—it can be used to manufacture adhesives, linoleum, varnishes, chewing gum, and printing inks; to prepare driers, resins, protective coatings, and corrosion inhibitors; and to make soaps, lipsticks, perfumes, shampoos, and sunscreens. Jojoba oil is also an excellent lubricant, and still other uses are being explored.

The list of benefits that species contribute to humans is already a

long one, but it can be lengthened even more. Most plants and animals do not now provide direct economic benefits to humans, but this
does not mean that those species will never do so.[36] Of the known
species of plants and animals, only a small fraction have been examined to determine their potential for new food and drugs or commercial and industrial products. About 10 percent of all plant species
contain substances that might be useful in treating cancers, yet few
of these species have been examined to assess their potential.[37] Of the
approximately 80,000 edible plants, humans have used less than 4
percent, and less than 200 are widely cultivated. A mere seven species
—corn, rice, wheat, barley, cassava, potato, and sweet potato—provide three-quarters of all human nutrition.[38] Almost all protein (from
domesticated animal species) that humans consume comes from just
nine species. Cows and pigs alone provide more than half of all meat
production. At the least, this extraordinary reliance on so few species
creates a high vulnerability to pests and disease. More important, limited diversity narrows the genetic base, thus reducing opportunities
to respond to these pests and diseases.

Wild species can also serve as indicators of environmental quality
as when lichens provide sensitive barometers of air pollution. For
millions of people, plants and animals offer symbolic, esthetic, and
recreational benefits. Most Americans take pride in the bald eagle as
a national symbol while many others enjoy the opportunities to observe flora and fauna in their native habitats. Many national opinion
surveys have demonstrated Americans' keen interest in wildlife. In
one such survey of Americans over age sixteen, the Department of the
Interior found that half of the respondents reported some "noncon-
sumptive use" of wildlife. In addition, the survey found that Americans had taken trips totaling nearly 400 million days to observe, feed,
or photograph wildlife.[39]

Every time a human contributes to a species' extinction, a range
of choices and opportunities is either eliminated or diminished. The
demise of the last pupfish might have appeared inconsequential, but
the eradication of other species could mean that an undiscovered cure
for some cancers has been carelessly discarded. The extinction of a
small bird, an innocent amphibian, or an unappealing plant might
disrupt an ecosystem, increase the incidence and areal distribution
of a disease, preclude the discovery of new industrial products, prevent the natural recycling of some wastes, or destroy a source of easily
grown and readily available food. By way of analogy, the anthropo-

genic extinction of a plant or animal can be compared to the senseless destruction of a priceless Renaissance painting or to the burning of an irreplaceable book that has never been opened. In an era when many people believe that limits to development are being tested or even breached, can humans afford to risk an expendable future, to squander the infinite potential that species offer, and to waste nature's ability and willingness to provide inexpensive solutions to many of humankind's problems? Many scientists do not believe so, and they are fearful of the consequences of anthropogenic extinctions. These scientists quickly admit their ignorance of the biological consequences of most individual extinctions, but widespread agreement exists that massive anthropogenic extinctions can bring catastrophic results.

In fact, when compared to all other environmental problems, human-caused extinctions are likely to be of far greater concern. Extinction is the permanent destruction of unique life forms and the only irreversible ecological change that humans can cause. No matter what the effort or sincerity of intentions, extinct species can never be replaced. "From the standpoint of permanent despoliation of the planet," Norman Myers observes, no other form of environmental degradation "is anywhere so significant as the fallout of species."[40] Harvard biologist Edward O. Wilson is less modest in assessing the relative consequences of human-caused extinctions. To Wilson, the worst thing that will happen to earth is not economic collapse, the depletion of energy supplies, or even nuclear war. As frightful as these events might be, Wilson reasons that they can "be repaired within a few generations. The one process ongoing . . . that will take millions of years to correct is the loss of genetic and species diversity by destruction of natural habitats."[41] David Ehrenfeld succinctly summarizes the problem and the need for a solution: "We are masters of extermination, yet creation is beyond our powers. . . . Complacency in the face of this terrible dilemma is inexcusable."[42] Ehrenfeld wrote these words in the early 1970s. Were he to write today he would likely add a note of dire urgency. If scientists are correct in their assessments of current extinctions and reasonably confident about extinction rates in the near future, then a concerted and effective response to human-caused extinctions is essential. The chapters that follow evaluate that response in the United States.

Possible responses to the problems of endangerment and extinction are large in number. Captive breeding, the establishment of refuges, the protection of habitats, and restrictions on hunting, development, and the illicit trading of endangered species represent only a few of many alternatives. Despite the range of possibilities, these alternatives have not been applied either sufficiently or successfully to halt widespread anthropogenic extinctions or to heal the wounds of most species on the brink of extinction. Christopher Stone offers a striking explanation for why this situation exists. Although his concern is the social control of corporate behavior, Stone's observation readily applies to the protection of biological diversity. In his words, "Nothing in society is a continuing problem because of itself, per se; something becomes and remains a problem because of shortcomings in the institutional arrangements we rely on to deal with it."[1]

The relevance of Stone's observation is underscored when one realizes that premature extinctions are infrequently due to biological causes. Most species are now endangered or become so because of actions that political or economic systems either encourage or allow people to take. This is particularly so because biological diversity represents a collective good, which no one owns since it is in the public domain, but which is readily accessible to anyone who wishes to exploit or benefit from this diversity. For such collective goods, individuals have no incentive to limit their consumption as long as other people also have unfettered access to these goods. The result is a tragedy of the commons in which pursuit of individual gain engulfs and overwhelms concern for the common good.[2] The only effective solution to the tragedy is collective action, which only governments can provide.

The issue is not only how much government intervention is desirable or necessary, but also how the governmental system perceives biological diversity as a political issue. However much scientists decry the causes and lament the consequences of human-caused extinctions, these scientists will be little more than concerned bystanders when policies are devised for the problems of endangerment. Scientific data about endangerment influences decision-making processes, but these processes are politicized ones occurring in a political arena. Oran Young indicates exactly what is likely to occur when Congress invites experts to testify. Young says that experts are "typically called in more for preconceived purposes of advocacy or negotiation than as sources of informed analysis and judgment."[3] Other evidence supports Young's assessment, and this provides further reason to examine the political arena and how it perceives issues. Such an examination, combined with an assessment of some theories about policymaking, can provide a sense of whether the American political system is likely to devote as much attention to the issue of endangerment as many scientists believe to be necessary. Equally important, the examination establishes a context of expectations in which to evaluate the prospects for the successful protection of biological diversity in the United States.

Popular Political Issues and Biological Diversity

Public officials make daily choices about how to use their time and which issues will benefit from their attention. These choices are related to the construction of political agendas, which include the list of topics or issues to which government policymakers are devoting serious attention.[4] Among all issues, far more candidates for attention exist than policymakers have time, energy, or incentive to consider. Many issues that seemingly deserve careful consideration are neglected and gradually fade in perceived importance. In contrast, other issues are lavished with attention, but such attention is not randomly bestowed. Policymakers prefer recurring, easily understood issues with which they have some experience and familiarity rather than new and complex issues with which they have little or no experience. This means that national political agendas are filled with such issues as taxes, budgets, national defense, and foreign policy.

An important consideration in the construction of agendas involves the potential political advantages associated with selecting new items for attention. The desire for reelection or the opportunity for political gain powerfully motivates many politicians, and this often leads them to favor or avoid issues that affect their continued electoral or political success.[5] The fear of adverse public reaction also keeps some issues, however meritorious, off political agendas. At least one caveat exists in regard to the perceived appeal of different issues: policymakers are not likely to raise an issue to agenda status *unless* they also have a favored and feasible solution to attach to the issue or problem. "In the absence of readily available solutions," explain two experts on political agendas, "we simply accept or learn to live with many difficult situations."[6] Available solutions allow policymakers to demonstrate their effectiveness to constituents and to be in position to propose new laws, which are often considered marks of political accomplishment.

What this means is that self-interest motivates most political behavior. One way to advance this self-interest is to identify a scapegoat or unpopular evil to be attacked and blamed for causing a problem. These evils usually appear in the form of industries or corporations that allegedly engage in disreputable behavior and that can be linked to the undesirable consequences. According to James Q. Wilson, "assigning blame becomes virtually a political necessity in order to get the issue on the public agenda" when the goal is to alleviate a widely distributed burden like pollution. For environmental problems like air and water pollution, it is frequently possible to identify, blame, and, if desired, penalize the culprits who caused the problem, just as Wilson believes to be necessary. He adds that this identification process is important in politics because it guarantees publicity and creates political momentum "in the competition for attention."[7] In contrast to many traditional environmental problems, concern for biological diversity faces a key barrier in this competition. When it comes to endangerment, corporate culprits are not easily identified, and establishing a causal linkage between an activity and a species' endangerment is often problematic. Other than identifying today's life-styles and Americans' preference for economic development, only occasionally is it possible to specify who is to blame for species' endangerment. As some scientists have complained, the "terrifying opponent against which conservationist forces can be rallied" is not available.[8] This understandably works to the disadvantage of those who demand effective policies to protect biological diversity.

The ability to assign blame is but one feature that affects the construction of agendas and policymakers' response to items on these agendas. Characteristics of the issues themselves are similarly important. Preferred issues include those that directly and tangibly affect large numbers of politically astute people who: a) consider a situation to be a problem in need of a political solution; b) believe they suffer or are disadvantaged because of the problem; and, c) demand a response from the political system. People or, more effectively, well-organized interest groups persistently lobbying are often able to influence the design of political agendas, either by forcing new issues before policymakers or, more likely, by blocking legislative initiatives and thereby maintaining the status quo.[9] With the latter situation more prevalent, new issues, even those including the most serious conditions, encounter problems in their quest for sustained attention.

This quest for attention from interest groups represents another area in which biological diversity finds itself handicapped relative to other political issues. The issue of endangered species has achieved agenda status in the United States, but its prominence and relative priority on this agenda is not without problems. It is almost axiomatic that members of large groups rarely organize to gain imprecise, collective benefits, such as widespread preservation of species. Members are widely scattered and have "no sense of special identity, no established pattern of interaction" and "no common interest in paying the cost of providing [a] collective good."[10] In short, although the benefits of preservation are highly desirable, political logic argues against the creation of powerful interest groups with concern for biological diversity. Groups interested in protecting this diversity do exist, but most focus narrowly on charismatic species like pandas or whales.

Interest group activity is not always a prerequisite for attention to an issue. Attention is assured in the face of crisis or disaster as, for example, when Iranians overran the American embassy in Tehran or when the Three Mile Island nuclear reactor nearly suffered a catastrophic meltdown. As John W. Kingdon observes in a study of political agendas, politics is all about crisis: "In the American system you have to get hit on the side of the head before you do something."[11] What makes a crisis or disaster, however, is a matter of interpretation.

An airplane crash that kills three hundred people is a disaster because it occurs infrequently; thousands of deaths due to separate automobile accidents over several months in many states are not viewed in a similar manner.

Disasters attract attention, but such is not always the fate of situations or problems considered to be routine and believed to have less immediate consequences. In fact, pressures to address today's problems (and fear of criticism for not doing so) mean that issues associated with anticipated or predicted harms that *might* occur some or many years into the future are likely to be neglected or denied a sense of urgency, regardless of scientists' pleas to the contrary.[12]

A sense of urgency about extinction may exist in the minds of many scientists, but this concern is not yet widely shared in the political community or among the general public. People may believe that there are undesirable consequences associated with premature extinctions. In the absence of a direct impact on life-styles, however, it is unreasonable to expect people to be too concerned with the endangerment or extinction of obscure plants or inconspicuous animals. People who live with the consequences of extinction, as millions of Americans do every day, will not consider it a matter that deserves their immediate attention.[13]

The problem is that scientists have not communicated successfully the adverse effects of past extinctions. This inability places scientists at a distinct disadvantage in their quest for effective political attention. Political systems place the burden of proof on those who want to change the status quo. Those favoring change must produce persuasive and occasionally even incontrovertible evidence that a problem exists. Circumstantial evidence or informed judgment is rarely satisfactory, especially when change threatens the well-being of those who profit from the status quo.

Scientists concerned about a crisis of extinction are further handicapped because they are woefully uncertain of the likely consequences of projected extinctions.[14] Under the best of circumstances it is difficult to communicate whatever information is available. The public and policymakers alike have a natural preference for simple and easily understood issues. Unfortunately, however, explaining the biological consequences of human-caused extinctions to an ecologically illiterate population is difficult. Even among those people who have made the most commendable efforts, agreement exists that the issues are "as complex and little understood as any that modern society has ever faced."[15] Informing the public is further hindered because, unlike other environmental areas, there have not been any newsworthy disasters involving endangered species *and* humans. Still another problem is that scientists and conservationists have been warning of an impending crisis of biological diversity for at least twenty years. The warnings may be justified, but people tend to neglect red flags that are raised and waved too frequently. When people are told that a biological crisis is occurring and then nothing seems to happen, scientists lose their credibility and are labeled as eccentric, publicity-seeking doomsayers.

Comparing other environmental problems with the issue of endangerment provides further support for these observations. Unlike the extinction-related consequences already experienced, the harms associated with hazardous wastes and air and water pollution are readily evident and directly affect those exposed to the pollutants. Despite these attributes, the United States still has major problems with these sources of pollution. The urgency to eliminate them is simply not evident or sufficiently persuasive to compel effective remedial action, and this is even more true for the issue of endangerment. For traditional environmental problems, moreover, the goal is to eliminate or control a known and immediate evil, a task likely to generate more support than the prevention of a potential or poorly understood future evil, as is the case with the consequences of the massive extinctions projected to occur. Cast in other terms, people are more likely to act to avoid an immediate threat in which they have a stake than they are to avoid some uncertain future risk, such as that associated with extinction.[16]

In short, many variables determine which issues catch policymakers' attention. Though this initial hurdle may be overcome, it does not ensure an effective response. Most agendas contain more items than can be acted on, especially because many items require regular attention, such as budgets. Some issues thus demand attention and action; their neglect is politically and administratively impossible. Other issues have certain features that have inherent appeal and thus create incentives for action. Among the many kinds of policies that can be formulated, the most popular are the ones that distribute material benefits to a majority or to a politically significant and effectively organized minority.

Perhaps no better example exists than the elderly in America, a force that few politicians can safely ignore. Americans between ages

55 and 64 vote at levels well above the national average, and the American Association of Retired Persons has nearly thirty million members, who produce annual revenues approaching $200 million. Much of this money is used for lobbying, and few people doubt its effectiveness. Compared to groups that are not well organized and that vote much less frequently, the elderly do quite well. Elderly Americans receive about ten times as much federal aid per capita as do children.[17] Furthermore, social security benefits are indexed for inflation and sometimes increased beyond the rate of inflation. In contrast, payments to poor families through the Aid to Families with Dependent Children program are not indexed, and the purchasing power of these payments declined substantially in the 1970s and 1980s.

Poor children and their families do not fare well in the distribution of public benefits, but at least they have a potential opportunity to influence public officials. Such is not the case for endangered species. They cannot vote, file grievances, organize protests, select representatives to lobby on their behalf, or otherwise mobilize themselves politically. Endangered species further have no financial resources, and they lack status, cohesion, and leadership, all of which affect a group's political clout. Just the opposite is true for the many economic interests that oppose policies designed to protect endangered species and their habitats.

Another feature of desirable policies is that they coincide with prevailing public opinion and do not require law-abiding people to change their life-styles or cause them inconvenience. Rather than banning the sale of tobacco, as an example, public officials focus on finding cures for tobacco-related illnesses. Occasionally public programs require individuals to change their behaviors, but such programs (especially among environmental policies) are rare and usually offer some subsidies to gain compliance.

Here is still another area in which the protection of biological diversity encounters problems as a political issue. If individuals' activities contribute to endangerment, then changes in these activities are likely to be part of an overall solution. The difficulty here is that the behavior of an entire population must be addressed rather than the actions of a single profit-seeking industry or otherwise limited group. One can hypothesize that the larger the group whose behavior is to be changed, the less likely the desired change will occur.[18]

One way to encourage behavioral change to protect biological diversity is through education. Unfortunately, educational campaigns to increase awareness of extinction-related issues are notably ineffective, perhaps because they are so complex. For the informed citizen who wishes to act responsibly, Myers is even more pessimistic: "A citizen is left bewildered, if not discouraged, by his inability to do much as a single citizen."[19] One of Myers's solutions is collective action, but as noted previously, collective action to obtain diffused and uncertain benefits is highly unusual.

When policies do call for change, those that have the best chance for success will rely on modest changes in current practices to address immediate, proximate causes rather than comprehensive changes in deeply embedded social behavior.[20] Recommending overhaul of these practices implies a condemnation of them as well as criticism of the people responsible for the practices. As a rule, the greater the change required, the less friendly the reception to the proposed change and the lower the probability of success.[21] When advocates of protection ignore this rule they increase the odds that their recommendations will be rejected.

Despite the plain logic of this maxim, it is breached more than observed, at least among many conservationists calling for greater efforts to preserve endangered species. Many scientists emphasize that only changes of extraordinary magnitude will be capable of solving the crisis of extinction. In their book on extinction, for example, Paul and Anne Ehrlich claim that "the problem . . . will not be solved by minor adjustments of the sociopolitical system."[22] Instead, the Ehrlichs declare a need for a universal transformation of society to include *halting* human population growth as quickly as possible and a complete restructuring of the economic system. The Ehrlichs may be correct, but political systems operate on the basis of political feasibility, not ecological necessity.

Preferred policies have other common features. The time span of elected officials is characteristically limited, and they often suffer a form of political myopia. In order to get reelected, these officials try to show that they have compiled an attractive record of accomplishment. Pressures to do so are evident for first-term presidents, but are especially intense for members of the House of Representatives, with their two-year terms. Facing an electorate accustomed to being told that its political representatives are diligently working on its behalf, presidents and members of Congress alike are ill-advised to ask constituents to wait many years to taste the uncertain fruits of an incumbent's electoral success. The usual result of this pressure is a preference for programs that provide concrete benefits in the short term, preferably as soon as possible.

This preference provides still another instance in which the protection of biological diversity suffers as a political issue. Some species can find themselves on the brink of extinction in the twinkling of an eye, but the period necessary to bring about their recovery is often measured in decades. Recovery efforts for whooping cranes, which started at least in the 1950s, continue today. Progress is evident, but the crane's situation is likely to entitle them to the endangered label for decades. Recovery efforts for the California condor will continue well into the next century, long after most readers of this book have died. The point is that few endangered species offer opportunities for the quick results that appeal to those with concerns for their next electoral campaign.

The incidence and distribution of costs and benefits also affect perceptions about the desirability of various programs. Except in unusual circumstances, the anticipated benefits of a program are expected to exceed its costs. One result of this preference is that easily monetized values, such as goods traded in the marketplace or development that provides jobs, will have advantages over values that are elusive or difficult to measure. It is much easier to estimate the economic benefits of building a coal-fired power plant than it is to assess the (potential) economic value of an apparently unremarkable insect or amphibian that might be exterminated because of the power plant.

The benefits of the power plant are appealing because the beneficiaries are easily identified, large in number, and the flow of benefits begins before construction. In contrast, many of the power plant's environmental costs are more difficult to assess because they will not occur or cannot be identified until some time in the future. As the U.S. budget deficit further testifies, deferring the costs of many programs to the future is exceedingly popular, especially when today's voters are the beneficiaries. In other words, a typical pattern of many public policies is to provide benefits now while imposing the costs of the policies on future generations. When costs are assessed, they are ideally diffused geographically while benefits are preferably concentrated among politically relevant groups.[23]

Conventional wisdom appears to argue for policies that provide concentrated benefits today while deferring and diffusing costs. Policies for the protection of biological diversity are not likely to fit this mold: preventing extinction requires that costs be imposed now in order to provide anticipated benefits at some later date. If collective action is the remedy, taxpayers will bear many of these costs. When it comes to many endangered species and recovery efforts, however, the costs will be highly concentrated. Some people may not be allowed to develop land they own because it provides the only known habitat of a nearly extinct plant. It may be necessary to prohibit fishing in some rivers so that rare fishes can be preserved. Commercial fishermen may find their operating costs increased because of the need to buy special nets to prevent harm to sea turtles. Although these costs will be concentrated, whatever benefits occur will be diffused widely and may not become evident until so many years into the future that their present value will be heavily or completely discounted.

The pattern of concentrated costs and diffused benefits is not unusual for environmental programs, but in this regard policies for biological diversity are likely to have at least one distinctive feature. Air and water quality programs attempt to reduce or eliminate what are familiar and proximate problems. In comparison, the goal of preservation programs is to prevent a speculative and poorly understood evil. This difference becomes important because many people do not appreciate the absence of evils that fail to occur. In addition, as Roy Gregory suggests, "no politician would bank on getting credit for negative blessings of this kind, even from those who have benefited the most."[24]

Finally, preferred policies typically offer supply-oriented solutions that emphasize the expansion of supply or the application of new technologies to solve problems.[25] Technological optimists abound in the United States, and inventions are seen as solutions to many problems, most prominently environmental ones. The legislative sponsors of the nation's clean air and water laws assume that American creativity and technological competence will ensure success. The laws are peppered with references to the mandatory application of the best available control technologies.

In addition to technology, many Americans also assume that expansion of supply offers a viable solution to their problems. In a country in which unbridled growth is assumed to be an unquestioned article of faith, reliance on supply-oriented solutions is not surprising. Few American policymakers have any experience with widespread or long-term scarcity. For the few with some relevant experience, most have been taught the rude lesson that governments are expected to provide solutions, not request the public to bear the consequences of shortages. Susan Welch and Robert Miewald summarize the situation well: "Alarming the public about a potential shortfall and possible future hardship is not a strategy perceived to win votes. . . . [F]ew elected politicians want to suggest solutions beyond finding more of the resource."[26] Welch and Miewald might also have recognized a related response, namely finding an alternative to the resource. As some economists have suggested, human ingenuity can provide suitable substitutes for whatever natural resources are exhausted.[27] The political appeal of supply-oriented solutions is undeniable. One of their key advocates occupied the White House during most of the 1980s.

Whatever the appeal of a supply-oriented approach, the wisdom of relying on it to prevent the loss of biological diversity is open to debate. This approach rarely addresses the root causes of a problem and also assumes that once a problem develops, it can be remedied without major changes in long-standing industrial or economic habits. Its ultimate presumption is, of course, that nature is incapable of imposing limits on human activities. When the proponents of an expanded supply offer an alternative, they usually argue for limits on government regulation and increased reliance on market forces to find the resource in short supply. The assumption here is that more of the resource (or a suitable substitute) is always available.

Some endangered species will undoubtedly benefit from new technologies, but one's optimism would have to be unlimited to expect technology to conquer the problem of anthropogenic extinction. Evolutionary processes occurring over millions of years are not easily duplicated, and technology is incapable of replacing something for which no models exist.[28]

This discussion points to the conclusion that endangered species do not now possess the attractive features to make them prime candidates for sustained political attention. Although many scientists are concerned about the endangerment and extinctions now occurring, their apprehension has not yet led to the changes these scientists believe to be imperative. In fact, the discussion suggests that many policymakers view the protection of biological diversity as a routine issue amenable to traditional, incremental solutions rather than as a distinctive issue requiring innovative responses. Support for this view comes from a congressional staff aide who attended a conference on biological diversity. After listening to three days of discussion about the likely loss of species and genetic diversity, the aide allowed that the problems are "tremendously important." Compared to other issues, however, the aide pointed out that the problems of extinction are "barely discernible" to Congress: "They form a small dot."[29]

If this assessment is correct and if scientists' projections of future extinctions are plausible, then the government agencies responsible for protecting and recovering endangered species face a discouraging scenario. The magnitude of their task is considerable, but their needs for resources, authority, and political capital may quickly exceed the political system's desire, capacity, and willingness to provide them. Much of what follows in the remainder of this book provides support for this view.

A Framework for Evaluation

Having discussed the issues of endangerment and extinction from both a scientific and political perspective, it is now possible to consider what has actually happened with the protection of biological diversity in the United States. Attention is directed at the various versions of the Endangered Species Act, and on the two federal agencies, the Department of the Interior's Fish and Wildlife Service (FWS) and the Department of Commerce's National Marine Fisheries Service (NMFS), that have primary responsibility for endangered species. The two agencies decide which species will be listed as endangered, which of their habitats should be protected, and develop plans for the recovery of species on the brink of extinction. The FWS is the more active of the two agencies and has management responsibility for all but a handful of endangered species, so it is the subject of most of the book's discussion.

In evaluating the services' efforts to protect endangered species, the key question is both brief and simple: How well do these efforts meet the purposes for which they were established? To answer this question properly, it is necessary to consider whether policies have been implemented as intended (an evaluation of process) and to assess how effective they are in producing the changes desired in the status of endangered species (an evaluation of outcomes and impacts). A related, but no less important question asks why these policies have or have not achieved their goals.[30] The emphasis throughout is on evaluation and applied policy research. Such research begins with a societal problem formulated outside the discipline, in contrast to basic policy research, which attempts to test general theoretical propositions about the causes or consequences of public policy.[31] The former focuses on the variables that governments can manipulate, rather than all variables relevant to an explanation, so applied research normally does not allow for the development and testing of general theories about public policy.

Identifying Program Goals

The first step in any evaluation involves a determination of goals or objectives. Clear, precise, and easily evaluated goals are desirable, but this message is rarely heard in the halls of Congress. The initial endangered species law provided one of many examples. The Endangered Species Preservation Act of 1966 was intended "to provide a program for the conservation, protection, restoration, and propagation of selected species." Federal agencies were directed to protect these species and, insofar as was "practicable and consistent with the primary purposes" of these agencies, to preserve the habitats of endangered species on lands under their jurisdiction.[32] This was an ambiguous declaration because it provided substantial opportunities for agencies to do exactly what they wanted and to act in their own self-interest.

This self-interest was addressed in subsequent legislation, the Endangered Species Act of 1973. This law, and later amendments to it, is intended to protect endangered and threatened species and to provide a "means whereby the ecosystems upon which endangered species and threatened species depend may be conserved." The act defines "conservation" to include the use of "all methods and procedures which are necessary to bring any endangered species or threatened species to the point at which the measures provided pursuant to [the] Act are no longer necessary."[33] In this instance, the goal is clear and unambiguous—the recovery of all species threatened with extinction.

For some evaluators an examination of statutory goals provides a sufficient statement of objectives as well as a logical place to begin an evaluation. Such a choice is unduly restrictive, however, because it ignores the dynamic nature of policy and the fact that bureaucrats and bureaucracies have goals that are distinct from and independent of the ones that laws impose on them.[34] An organization's goals can include survival, autonomy, expanded jurisdictions, training and promotion of personnel, and friendly relations with legislators and other agencies. Moreover, all agencies are subject to real or imagined pressures from constituents, interest groups, Congress, the courts, the executive, other administrative agencies, and state, local, and even foreign governments. As a result, the goals of any public policy are best examined in the context of the characteristics and the operating environment of the implementing agency. In other words, organizational goals and environments shape, mold, influence, and affect how agencies respond to the goals found in statutes. These considerations are the subject of chapter 3.

Construction of an Impact Model

Once a policy goal has been defined, the next step is to identify the activities needed to achieve that goal. This process is one of constructing an impact model that outlines a sequence of causally linked activities from input to outcome and eventually achievement of the ultimate goal. An ideal impact model indicates not only the activities that must be undertaken but also how successful accomplishment of the related intermediate goals contributes to success in reaching the ultimate goal.[35] The explicit assumption of all impact models is that accomplishment of intermediate goals will cause or produce achievement of the ultimate goal within a specified time period.

Successfully reaching intermediate goals does not always lead to ultimate goals. Some impact models are invalid; in such a case, failure to achieve ultimate goals is due to a faulty impact model, not to faulty implementation. Helen Ingram describes the possible consequences of a faulty impact model as "doing the inappropriate thing very effectively."[36] Consequently, construction of an impact model and examination of its validity helps to identify the causes of a policy's relative success or failure.

In its deliberations on the endangered species laws, Congress has never specified an explicit impact model, thus frustrating systematic evaluation and, possibly, the likely achievement of its ultimate goal. Nonetheless, the major intermediate goals include:

1. The listing of species in jeopardy of extinction
2. The protection of habitats
3. The protection of listed species through enforcement of regulations governing activities directly affecting such species and their habitats

4. Interagency consultation to prevent harm to listed species or their habitats, and
5. The implementation of recovery plans

Taken together these goals provide only a semblance of a well-explicated impact model. The endangered species laws do not contain an explanation of how the intermediate goals are related, which are the most important, or any indication of how much of each activity —listing, protecting habitats, implementing recovery plans—is necessary to ensure ultimate success. Despite these problems, Congress assumes that the activities make a positive contribution to the recovery of endangered species, so chapters 5 through 10 evaluate how well the intermediate goals have been achieved and with what degree of effectiveness.

Development of a Research Design

Textbooks abound with recommended designs for program evaluation. Experimental designs, with random assignment to treatment and control groups, are always desirable. So-called quasi-experimental designs, used when randomization or comparison is not possible, are less desirable, but are still on the preferred list of designs. In either case, an appropriate design permits at least three things. First, the evaluator can reliably compare a situation before and after a treatment (such as providing methadone to a heroin addict) or a policy intervention (such as a program to reduce unemployment). Second, an appropriate design allows a determination of whether any change that does occur in a treatment group is due to the intervention rather than to other, extraneous variables. Last, a proper design allows one to generalize from the sample included in the evaluation to the larger population from which the sample is drawn.

Despite their appeal, experimental and quasi-experimental designs are often difficult to apply in the evaluation of public policies, including the preservation of endangered species. There is only one Endangered Species Act, so alternative approaches to protection are not easily examined. No set of endangered species is representative of all such species, so for purposes of comparison it is not possible to use a control group (i.e., a group that does not receive the treatment, in this case the alleged benefits of protection).

Furthermore, preferred designs incorporate certain assumptions that are not readily applicable to endangered species. First, for example, in order to assess the consequences of a public policy, it is desirable to know the status of the situation or problem before the intervention begins. For endangered species it is frequently difficult to be sure of any species' status. A naturally rare species could be increasing in numbers, yet still be considered in danger of extinction while a relatively healthy population of another species could be experiencing a temporary, but not fatal decrease in its numbers. Conservationists could easily misinterpret either situation and reach unjustified conclusions about a species' status.

Second, the use of before-and-after measures inherent in preferred designs assumes that the target group (such as drug addicts or cities with high unemployment) receives the same or similar treatments over the same time period. When this occurs it is possible to evaluate the consequences of a single program or at least a set of closely related program elements. For endangered species, this assumption is not valid. The treatment provided to each endangered species differs in many ways. Furthermore, the purpose is not to evaluate a single law, but rather the comprehensive impact of a series of laws, the first of which was approved in 1966. Major revisions and amendments to this law occurred in 1969, 1973, 1978, 1982, and 1988. The consequence is that there is no single policy action, except listing, that all endangered species share in common.

These problems do not leave the evaluator without hope. The purpose of a research design is to guide the evaluator, not to prevent evaluation. Regardless of what kind of approach is used, the quest is for "systematic and empirical evidence of the extent to which the policy has accomplished its goals."[37] The intermediate goals included in the impact model offer guidance about the kind of data that should be collected. Moreover, the clarity of the ultimate goal makes it relatively easy to determine whether it has been achieved. Finally, the issue of whether relevant activities actually cause or bring about the recovery of endangered species becomes important only if recovery occurs. If there are few instances of recovery, this alone provides evidence of a faulty impact model, flawed implementation, or perhaps even insurmountable biological constraints that preclude success.

Data Collection

Few government agencies encourage systematic evaluation of their programs. Annual reports are typically filled with data relating measures of administrative effort and expenditures rather than measures of outcome.[38] Agencies have considerable control over the former. They can affect the number of people assigned to a program, report the number of clients served, the amount of money spent, and so on. Such data can be manipulated to cast a favorable light on just about any government program. In contrast, agencies typically have much less control over outcomes, which represent the "meat and potatoes" of a meaningful evaluation. The result is that the typical government agency publishes measures of accomplishment relatively infrequently. Furthermore, when these measures are reported, they usually address intermediate rather than ultimate goals.

The endangered species programs of the FWS and NMFS are not exceptions to these habits. The FWS regularly publishes the *Endangered Species Technical Bulletin*, which reports on recent protection-related activities. Every issue of the *Bulletin* includes information on the number of species listed as endangered without indicating whether the species' situation has improved because of its listing. The *Bulletin* likewise summarizes the number of recovery plans approved and the number of cooperative agreements the FWS has signed with the states. Readers are supposedly to believe that these administrative accomplishments beget biological success.

Accepting this belief requires more faith than any reasonable evaluator has, so it is necessary to go well beyond readily available information. In addition to the usual books and articles, data for the present study come from many sources, including a reading of thousands of pages of the *Federal Register* and congressional hearings and reports, correspondence and unpublished documents in the files of the FWS and NMFS, and interviews with FWS and NMFS employees as well as with state officials responsible for endangered species in Hawaii and California.

A comprehensive data set was also created for each native American species ever listed by the FWS and NMFS.[39] For each of nearly six hundred species of plants and animals, data were collected on such variables as the dates of proposed and final listing, the FWS's listing and recovery priorities, whether a species' habitat has been formally protected, whether it has a recovery plan, and on measures of the degree of threat and potential for recovery for each species. All these data serve as the basis for much of the analysis that follows.

Interpretation

After analysis it is time for conclusions about a program's overall effectiveness. In some evaluations this calls for a judgment about whether a program is a success. To make such a simple judgment is rarely possible and is usually undesirable. What constitutes success for any public policy is open to interpretation; subjective elements color any judgment.[40] Consider, for example, the nearly contemporaneous appraisals of the Endangered Species Act that representatives of two environmental organizations provided in the mid-1980s. In the first, a spokesman for Defenders of Wildlife applauded the act as "the most comprehensive and ambitious attempt any country has ever made to preserve wildlife." A few months later the Conservation Foundation concluded that "the nation's endangered species conservation program is woefully inadequate and is failing to prevent the impoverishment of much of our rich biological heritage."[41] Who is correct? Perhaps either, neither, or even both!

Judgment is further clouded once one appreciates the magnitude of the task and the constraints the FWS and NMFS encounter in trying to complete their appointed rounds. As Eugene Bardach has generalized, "Any social program worth having a governmental policy about . . . is likely to be a serious and complicated problem and therefore not amenable to easy solution or even amelioration."[42] This statement surely applies to the protection of biological diversity, which involves one of the world's most intractable problems. The Endangered Species Act establishes an ambitious goal of complete recovery for all species in jeopardy, but few knowledgeable people consider this to be an achievable goal. Many species are so intrinsically vulnerable to extinction because of their limited range, natural rarity, or unusual habitat requirements that it is unreasonable to anticipate their complete recovery.[43] If many species cannot be recovered, perhaps one ought to consider other evaluative criteria. Rather than using recovery as the criterion, an alternative goal might be the prevention of anthropogenic extinctions. Unfortunately for the FWS and the NMFS, even the prospects for achieving this more modest goal are likewise open to considerable doubt.[44]

One reason for this doubt is that opportunities to succeed are hindered. The Endangered Species Act does not include measures to prevent species from becoming endangered due to human activities. Prevention is the best cure for most ailments, but it is not one that is allowed in the medical kits of either the FWS or the NMFS. The agencies can act only after a problem arises when, as Michael Bean has noted, "the easy and inexpensive options have been lost."[45] This means that the two agencies do not have a fair chance to succeed; it is as if some-one has included hurdles in a three-legged marathon in which the two agencies are required to participate.

By way of comparison, imagine a situation in which the Environmental Protection Agency (EPA) knows what will cause an environmental catastrophe but then is not allowed to prevent or remedy the problem until someone has identified the victims, most of whom have already died. The few survivors have all been admitted to a hospital emergency room, but no one there knows for certain what treatment will ensure their recovery. To extend the comparison, the survivors need intensive and highly specialized long-term care. As long as the survivors remain hospitalized, any prognosis of complete recovery is premature. Recovery depends not only on the medical attention given to the patients but also to many things over which the doctors have limited control. The parallel point for endangered species is that a conclusive assessment of performance may likewise be premature. The task is enormous, much remains to be done, and success is not easily achieved, especially in the organizational environment in which the key governmental actor, the Fish and Wildlife Service, finds itself.

3 Administrative Capacity for Action

Administrative agencies clearly differ in their political clout. Some agencies, like the Federal Bureau of Investigation, the Corps of Engineers, and the National Aeronautics and Space Administration in the 1960s, are powerful and find that relatively few obstacles confound their administrative missions. Other agencies are weak, highly sensitive to external pressures, and apprehensive that their decisions will jeopardize whatever public and administrative support exists for their activities. "Some agencies are extraordinarily gifted in their ability to achieve their goals," notes Francis Rourke, while "others often seem to be stepchildren of the executive branch."[1]

Explanations for differences in administrative power are legion, but one explanation focuses on the attributes of the organizations. Differences in power are related to such political and organizational characteristics as an agency's mission, its size and resources, location within the bureaucracy, ability to provide tangible benefits, use of scientific or technical information, and the size and political skills of its clientele. The extent to which an agency must depend on the cooperation of other agencies also affects its actions. Each of these factors is useful in understanding Interior's implementation of its endangered species activities and, more precisely, the related activities of the FWS.

Institutional Location

A potential determinant of an agency's power is its location in the bureaucratic maze. Due to their stature and institutional location, cabinet departments and independent regulatory commissions are relatively autonomous, have significant influence and political clout, ready access to key decisionmakers, and considerable control over the design of their budgets and programs.

In contrast, organizations can be placed deep in the administrative labyrinth, far from the centers of political power and subject to many burdensome layers of administrative control. These organizations are best described as administrative wallflowers, often lacking status, visibility, and the respect of other organizations. Organizational weakness is compounded when an administrative unit is one of many offices or bureaus within a single umbrella agency or cabinet department that has many responsibilities. Such is the unfortunate fate of the FWS, which one of its superiors once described as "merely a little branch buried in the Department of the Interior."[2]

With the passage of the Endangered Species Act and the Fish and Wildlife Coordination Act, such a conclusion is no longer as applicable as it once might have been. Despite sizable increases in its responsibilities, however, the FWS remains deep within the bowels of the Interior Department. Interior is a conglomeration of four bureaus (the Bureaus of Mines, Reclamation, Indian Affairs, and Land Management), three services (the FWS, the NPS, and the Minerals Management Service), the U.S. Geological Survey, the Office of Surface Mining Reclamation and Enforcement, and an office for territorial and international affairs. Indeed, the FWS represents only a small part of Interior, as the data in table 3.1 show. The number of its employees ranks the FWS as the sixth largest unit in the department. Within the department, however, the FWS is paired with the NPS, Interior's largest unit, in a reporting relationship to one of five assistant secretaries, the assistant secretary for fish and wildlife and parks. The result is that approximately 70 percent of this assistant secretary's personnel are in the NPS.

Dissecting Interior provides an incomplete picture of the Service's organizational status and institutional location. The Service's institutional parentage is also of interest. Its genealogy began in 1871 when Congress required the president to appoint a commissioner of fishes and fisheries. Congress expressed additional interest in fisheries in 1903 when it created the Bureau of Fisheries and placed it in what was then the Department of Commerce and Labor. In turn, the new bureau assimilated the responsibilities of the fish commissioner, who had been located in the Department of the Treasury. The bureau

Table 3.1   Department of the Interior Enacted Budget Outlays and
Full-Time Equivalent Staffing, FY 1990

Bureau or Office	Budget Outlays (in millions)	Percentage of Total	Staffing	Percentage of Total
National Park Service	$1,057	13.5	17,256	23.9
Indian Affairs	1,715	21.8	12,478	17.3
Land Management	1,064	13.6	11,348	15.7
Geological Survey	468	6.0	9,705	13.4
Reclamation	1,092	13.9	7,206	10.0
Fish and Wildlife	855	10.9	7,007	9.7
Mines	171	2.2	2,342	3.2
Minerals Management	634	8.1	2,076	2.9
Surface Mining	303	3.9	1,200	1.7
Other[a]	489	6.2	1,685	2.3
Totals	$7,848	100.1	72,303	100.1

Source: Department of the Interior, *The Interior Budget in Brief: Fiscal Year 1991 Highlights* (Washington, 1990), 70, 118–229.
[a]Includes Office of the Secretary, Office of the Solicitor, Office of Inspector General, Territorial and International Affairs, and the National Indian Gaming Commission.

was transferred to Interior in 1939 and, in the following year, combined with the Bureau of Biological Survey (which, until 1939, had been in the Department of Agriculture) to form the FWS. This marriage lasted only sixteen years, primarily because of complaints that the FWS favored research and the management of wildlife refuges at the expense of commercial fisheries.[3] To address these complaints the Fish and Wildlife Act of 1956 retained the FWS as an organizational entity but created two separate units within the Service, the Bureau of Commercial Fisheries (BCF) and the Bureau of Sport Fisheries and Wildlife (BSFW). The bureaus' goals were not entirely congruent. One favored development and consumptive uses of marine resources; the purpose of the BCF was "to help the U.S. fishing industry maintain the position of the United States as one of the world's leading fishing nations." The BSFW, while catering to recreational hunters and fishermen, was dedicated primarily to conservation. Regardless of the division of responsibilities, the BCF found itself uncomfortably located in the Interior Department and, in 1970, was transferred to the Department of Commerce, where it became the National Marine Fisheries Service (NMFS). The Bureau of Sport Fisheries and Wildlife became the FWS in 1974.

These frequent reorganizations reflect poorly on the Service's political support and administrative strength. Organizations have a natural aversion to change and tend to resist it, particularly when it is packaged as reorganization, which indicates some deficiency in existing organizational arrangements. It should not be surprising that reorganization typically demoralizes employees. As Jeanne Clarke and Daniel McCool cast it, "there is little doubt that from an agency's point of view the costs of reorganization outweigh its benefits."[4] They surely had the FWS in mind when they reached this conclusion. Among the seven agencies they examined, only the FWS has had such a checkered organizational history. Rourke's assessment is likewise negative.[5] He contends that frequent reorganization is indicative of limited support. In contrast, he notes, agencies that serve strong clientele groups are "able to resist having reorganization imposed on them."

If the FWS is seen as an incidental appendage buried in Interior, the same can be said about the service's Office of Endangered Species (OES). It was created in 1966 with a staff of two. The OES budget was a comparably small portion of the Service's total budget, and it represented less than one-half of one percent of Interior's total budgetary authority. Several hundred FWS employees now have responsibilities for endangered species, but this staff size still represents less than 10 percent of all FWS employees.

For much of its life the OES was one of three offices and one division reporting to one of five FWS associate directors. A reorganization in mid-1986 created twenty-three divisions within FWS. One of these, a Division of Endangered Species, became one of five divisions reporting to one of five FWS assistant directors. Still another reorganization occurred in late 1987, but it was more significant than all previous ones. The OES was eliminated as a separate office, and its responsibilities were divided among several branches in a new Division of Endangered Species and Habitat Conservation. Technically this reorganization represented an increase in rank and visibility for endangered species activities, but the change was not well received among many FWS employees. In an effort to improve efficiency nearly one-fourth of all OES positions were eliminated because of the reorganization.

In sum, institutional concern for endangered species is truly buried in Interior. Is it also the situation that the division finds itself in a department with competing or incongruent values? An examination of the missions of the Department of the Interior and the FWS provides some answers.

The Department of the Interior's Mission

Among all cabinet departments probably no other has as many diverse and competing responsibilities as does the Department of the Interior. Its mission is "to encourage and provide for the preservation, development, and management of the natural resources of the United States for their use, enjoyment and security of its people, now and in the future."[6] How well the department meets its dual mandate to preserve *and* develop is subject to frequent debate. William Wyant states the case well when he notes that Interior's federal land management agencies always face a conflict in their missions: "It is the conflict, no doubt also unavoidable, between the task of conserving for the future and the task of exploiting for present use. At the same moment one agency may play the defending angel while another, in the same building, perfumes the sheets for the ravager."[7]

The FWS and NPS are among the alleged angels while the prodevelopment forces traditionally include the Bureaus of Mines, Reclamation, and Land Management. According to Nathaniel Reed, an assistant secretary for fish and wildlife and parks in the 1970s, the three prodevelopment bureaus "advocate exploitation of natural resources and [tend] to view environmental safeguards as roadblocks in the way of progress."[8] With Interior's dual responsibilities, an important question focuses on whether both goals are achieved simultaneously or, alternatively, which goal receives preference. The answer is critical to the FWS's mandate to protect endangered species. If efforts are inconsistent with the values that prevail elsewhere in Interior, the FWS will receive neither a warm response to requests for support nor an enthusiastic welcome to its recommendations about endangered species. In other words, the Service's capacity for effective action is largely dependent on the organizational environment in which it operates.

A closer examination of Interior's tasks reveals a preference for development over environmental preservation. The difficulties in developing and preserving are especially evident when one realizes that Interior manages over one-third of *all* land in the United States and most of its federally owned natural resources. It is estimated that more than 80 percent of the undiscovered crude oil in the United States will be found on federal lands or at federally controlled offshore locations. Likewise, federal areas are believed to contain nearly all the tar sands, 80 percent of the oil shale, 40 percent of the natural gas, and 35 percent of the coal yet to be discovered and exploited in the

United States.[9] The nation's mineral wealth includes large amounts of uranium and untapped deposits of several minerals now imported, including cobalt, platinum, and chromium. Geothermal opportunities are also widespread on federal lands.

One of the department's tasks is to regulate the exploitation of these resources through leases for coal, oil shale, geothermal power, and on- and offshore deposits of oil and natural gas. The pressures to develop the energy resources are periodically intense, depending on the perceived severity of the current energy crisis and the views of the incumbent presidential administration. President Nixon, for example, proposed accelerated research on oil shale, increased dependence on coal from federal lands, and a significant increase in the leasing and development of the outer continental shelf. The desire to increase the exploitation of these energy resources was a bipartisan one. President Carter stressed a continuing increase in the production of offshore oil and new incentives to utilize coal and oil shale resources on federal lands.

The Reagan administration judged these initiatives to be inadequate, claiming that it had inherited a nation suffering from a lack of leadership in the development of its natural resources. Prior to 1981, according to Interior, "Our economic and national security were in jeopardy because of a failure to develop properly America's abundant natural resources."[10] To buttress its case, the department noted that no leases for the development of tar sands had been issued since 1965, no onshore oil or gas leases in Alaska since 1967, no oil shale leases since 1974, and that only 4 percent of the entire outer continental shelf had been made available for exploration of oil and gas. Reagan's Interior Department acted quickly to redress these alleged failings. By late 1983 the department had leased more than six times as much federal coal as in the previous seven years, nearly doubled the acres leased for onshore oil and gas development over the preceding three years, and had issued nearly 1,200 oil and gas leases covering more than four million acres in Alaska. Wildlife refuges were not immune from pressures to exploit publicly owned land. In late 1986 the FWS recommended that large portions of Alaska's Arctic National Wildlife Refuge be opened to oil and gas exploration, despite anticipated adverse consequences for the resident wildlife.

A further reason to lease federal lands for energy development relates to the economic incentives of such action. Interior is one of a few federal agencies in which revenues typically exceed expenditures.

Leases for oil and gas drilling alone generated rents, royalties, and bonuses exceeding $5.7 billion in fiscal year 1986, and some of this money was used to offset the department's budget requirements. Despite its large annual revenues, Interior typically receives much less money than it needs to meet its responsibilities. This occurs because much of the money it collects goes directly to the Treasury Department. William Wyant's views are not at all atypical of prevailing opinion:

> In the management of federal lands, the legislative branch has been niggardly. . . . A key sin of omission has been failure to give the land-managing agencies the moral support, men, and money they need for the proper discharge of a heavy responsibility. Far from being willing to provide the tools essential for good husbandry, Congress has frequently taken the attitude that federal officials trying to protect the common heritage are guilty of harassment. . . . Congress has [deliberately] . . . kept the federal land departments in an impoverished condition.[11]

The condition and usage of much publicly owned land in the West provide a further illustration of Interior's difficulties in preserving and developing in a balanced manner. Interior's Bureau of Land Management (BLM) is the federal government's largest landlord, managing more than 325 million acres. Nearly all this land is located in eleven western states and Alaska and has long been available for mining, logging, and grazing of privately owned livestock. Possessed of a narrow geographic constituency, and subject to the effective pressures of a narrow spectrum of special interests, the BLM is often cited as a leading example of regulatory capture, beholden to the developmental interests it is expected to control.[12] If, indeed, the BLM is an example of regulatory capture, legislative mandates provide an explanation for its alleged bias.

In response to substantial deterioration of public lands in the West in the late 1800s and early 1900s, Congress passed the Taylor Grazing Act in 1934. Its purpose was to oversee livestock grazing on the public lands, but in doing so it ignored all other potential uses for the land except grazing, which was declared to be the principal use of the land.[13] The law was intended to prevent overgrazing, but it has met only questionable success. After its passage much of the land continued to face severe overgrazing, albeit with the approval of powerful ranchers and influential members of Congress. The U.S. Grazing Service, established by the act and charged with the responsibility for monitoring much of the land used for grazing, initially tolerated the lands' deterioration through inadequate regulation and low fees that encouraged overuse. Subsequent suggestions to raise the ranchers' fees for use of this land were successfully opposed, as were efforts to prevent overgrazing. Eventually, however, the Grazing Service proposed more stringent enforcement of its regulations, an increased staff to manage its lands more effectively, and fees that realistically reflected the market value of the forage. The congressional response was both swift and sweeping. As one former BLM employee noted: "[A] sorry compromise was reached—Congress cut Grazing Service appropriations in half and reduced its personnel by two-thirds. Only 50 workers were left to run the grazing districts—one for each 3 million acres. The net benefit to the public was a tiny increase in the grazing fee."[14] The Grazing Service's skeletal remains did not long survive. It was combined with the BLM in 1946.

Since then the BLM has had only limited success in improving the quality of its rangeland. As late as 1975 the bureau reported that nearly 80 percent of the land it manages in eleven western states was in fair condition or worse. More recent studies reach similar conclusions, namely that BLM land management practices are harmful to other possible land uses, such as for wildlife and fisheries.[15] These findings are not surprising when the views of some BLM administrators are considered. One of these officials, the bureau's assistant director for renewable resources, once told an investigator from the General Accounting Office (GAO) that the agency's "wildlife program will lose if it doesn't bend to commodity production. The best wildlife can do is not impede commodity oriented production."[16]

Examples of BLM land management policies implemented with apparent disdain for wildlife are not limited to grazing practices. Agency policy regulating the mining of nonenergy minerals strongly encourages resource exploitation. Likewise, the BLM has also been the subject of severe criticism from the GAO, the OTA, and many public interest groups. These criticisms focus on the anticipated adverse environmental impacts associated with the coal-leasing program as well as on the bureau's proindustry bias in the administration of the program.[17]

Brief mention should also be made of the BOR, another FWS sibling. This bureau's mission is to develop and operate water resource projects in seventeen western states. These projects range from irrigation and flood control activities, municipal and industrial water supplies, to the generation of hydroelectic power. The bureau has unabashedly viewed its goals in a developmental perspective. In the opening paragraph of his book on the BOR, William Warne, a former assistant commissioner of reclamation, boasted that the bureau is "a twentieth-century product of the national policy to develop the Western United States. The federal reclamation program . . . has been one of the most effective tools for the development of the West." [18] The bureau's emphasis on development represents a quintessential example of a distributive agency in that it provides benefits to organized groups. These benefits, though limited geographically, foster exceptionally high clientele support for the bureau. [19] Its projects have included Grand Coulee and Hoover Dams, California's mammoth Central Valley Project, as well as large projects in the Colorado, Columbia, and Missouri River basins. These and other bureau projects have been criticized not only for their substantial subsidization of water users but also for the projects' damage to the environment.

One recent study attempted to put into a comparative perspective the relative environmental values of the BLM and the BOR. [20] Twenty mid- to upper-level officials in the two bureaus (and five other natural resource agencies) were asked to rank each of the seven agencies, except their own, in terms of the quality of the agencies' response to the National Environmental Policy Act (NEPA). The two bureaus were by far the lowest-ranked agencies. No respondent ranked either as the most responsive agency among the seven; indeed, most respondents ranked the bureaus as the least responsive to environmental concerns.

The FWS fared somewhat better in this ranking scheme, but it received a lower score than one might expect from an agency with its responsibilities. Its mission is "to conserve, protect, and enhance fish and wildlife and their habitats for the continuing benefit of the American people." [21] To accomplish this mission the FWS operates 70 fish hatcheries, 160 waterfowl production areas, and 450 national wildlife refuges. In spite of these duties, the FWS is not free of the conflicting pressures that characterize the mission of its parent, Interior. When Lynn Greenwalt testified during Senate hearings on his confirmation as FWS director in 1974, he offered a revealing statement of what he saw as the Service's obligation to balance its mandates. He noted that the Service's role is to act as both a steward and an advocate for fish and wildlife resources. "This is not to say," he added, "that these resources should not be developed to meet social and economic needs. Unquestionably, they should and must be developed." [22]

Several examples of FWS management of different resources also provide an improved understanding of the difficulties in balancing diverse mandates. Until 1986, the Service had responsibility for implementing the Animal Damage Control Act of 1931. This law authorized the FWS to "eradicate, suppress, destroy or bring under control predators" such as bears, bobcats, coyotes, mountain lions and, until 1964 and 1971 respectively, the red wolf and the eastern timber wolf, two species that are now in danger of extinction.[23] However opponents feel about the control program, it is a popular one among many western farmers and ranchers, and millions of animals have been killed or removed as part of the program. Congress has traditionally been generous in its support of the program, and it was a favored one during the Reagan years. When the administration proposed its initial budget revisions in early 1981, the FWS was slated for substantial reductions, with some activities losing as much as 40 percent of their funding. In contrast, additional money was requested for animal damage control.

Controversy occasionally envelops the program because of the methods used to kill predators. One method relies on Compound 1080, a poison developed to combat rodents.[24] Although highly effective in killing coyotes, some endangered species have also become victims of the poison. Opposition to Compound 1080 culminated in early 1972 when President Nixon issued an executive order banning its use and similar toxicants as a means of predator control on all federal lands. Later, the EPA suspended the poison's registration. However desirable the decision, the director of the FWS asked the EPA to reregister the poison in 1981. The following year President Reagan rescinded Nixon's executive order, thus allowing the use of specific predacides on federal lands.

The FWS similarly faces conflicts over the use of the nearly ninety million acres it manages. The FWS is the federal government's third largest land manager, with responsibility for more land than even the NPS. The FWS land has valuable assets other than fish and wildlife, and use of the wildlife refuges is not limited to the needs of fish and wildlife. The National Wildlife Refuge System Administration Act of 1966 authorizes the secretary of the interior to "permit the use of any area within the [Refuge] System for any purpose . . . whenever he determines that such uses are compatible with the major purposes for which such areas were established" and *do not* interfere with the dominant use for fish and wildlife. Economically productive uses can include farming, grazing, logging, mining, and commercial fishing and trapping. In addition to recreational hunting and fishing, other public uses include camping, boating, and the use of off-road vehicles.

These uses provoke controversy and claims that the Service's ability to protect fish and wildlife is being compromised. There is considerable evidence to support these claims. One survey of its refuge managers in the late 1980s found that most refuges suffer from the consequences of secondary, nonwildlife-related uses. More often than not, these uses are incompatible with the purposes for which the refuges were established. When asked to identify which secondary uses are most likely to be harmful to these purposes, the managers listed mining, off-road vehicles, recreational boating, and water skiing.[25] Many managers complained that these activities detracted from their primary wildlife-related responsibilities and diverted resources away from wildlife protection and enhancement.

Managers objected to such problems, but the managers indicated in the survey that they have little choice. Determinations of compatibility are supposed to be based on site-specific biological considerations, but nonbiological factors and external political pressures often prevail. One FWS official explained that in deciding whether to allow approval of rights-of-way across refuges, "the strongest biological reasons for disapproving the use can often be overcome by the weakest economic rationale."[26]

The FWS also encounters management problems because it is unable to provide enough people to staff or maintain all of its refuges or even to enforce relevant laws, including those for hunting. Consequently, the FWS finds itself dependent on cooperation from state wildlife agencies, which have many more people in the aggregate than does the FWS. What many of its state-counterpart agencies also have is a traditional orientation toward resident game species. The Service's dependence on the states creates opportunities for state and local preferences to receive favorable consideration.

One of the largest refuges in the continental United States, the Charles M. Russell National Wildlife Refuge in northeastern Montana, provides an illustration of this point. When the refuge was established in 1936, it had two primary functions: livestock grazing and protection of grouse and antelope. Responsibility for the functions was divided between the FWS and the BLM until 1976 when the Service was given full responsibility. With over 800,000 acres, the Russell Refuge is the largest grazing unit in the refuge system, and the range provides more in grazing fees that any other. One study found that much of the grazing is harmful to the refuge, with cattle competing with game animals for food and destroying nesting cover for birds.[27] Due to pressure from the state government and affected ranchers, the FWS resisted efforts to reduce the grazing, and in at least several respects encouraged it. Local ranchers successfully sued in federal district court to prevent a reduction in grazing on the range.

Overgrazing is a problem on some refuges, but it is not the only one. Most refuges suffer because of poaching, soil erosion, wildlife disturbance, off-road vehicles, or air and water pollution. Some of the degradation is irreversible. External sources, such as acid rain and pressures from development, cause most of the problems, but internal sources are not blameless. Due to inadequate controls within refuges, grazing, farming, logging, and the use of pesticides contribute to refuge degradation. An Interior report issued in 1985 identified 121 refuges suffering from water contaminated by oil spills, pesticides, fertilizers, saltwater intrusion, industrial wastewater, or toxic substances like dioxin, asbestos fiber, and even plutonium and mustard gas.[28] In perhaps the most egregious example of environmental degradation within the refuge system, Interior was forced to close the Kesterson National Wildlife Refuge in central California in the mid-1980s due to the presence of large amounts of selenium in the refuge's ponds and marshes. The selenium, which had been leached from soils, reached the refuge as a result of runoff from nearby farms and because the BOR decided in the mid-1970s to run a drainage canal into the Kesterson site. The consequences were quickly noted—most fish within the refuge died, snakes and frogs disappeared, and many birds exhibited birth defects.[29]

Not all pressures on the refuges or the FWS come from external sources. Because of its "preservationist" leanings, the Service is often criticized for being antagonistic to growth and development. Several secretaries of the interior have attempted to change this view. President Eisenhower's secretary, Douglas McKay, was an automobile dealer (as well as governor of Oregon) before his appointment. He opened refuges to oil and gas exploration, developed a reputation for being friendly with private interests, and eventually became known as "Giveaway McKay." His frequent response to problems reflected his occupational experience: "Well you know, I'm just a Chevrolet salesman."[30]

The FWS faced a salesman of a different kind when James Watt became secretary of the interior in 1981. He could not have made his disdain for the Service's perspective more clear than when he told a congressional hearing a few weeks after his confirmation as secretary that he wanted to make major changes in the management of the FWS. The Service, he said, "need not stop economic activity, economic growth and job opportunities. In too many instances I believe that it has. . . . With the changes in budgetary requirements that I am requesting, we will get the attention of the management of the Fish and Wildlife System and new ways and new techniques will be installed throughout the Service."[31]

The FWS did not neglect the secretary's admonition. The Service was soon required to devote more of its scarce resources to energy-related federal projects, and it worked to expedite the development of deepwater ports to facilitate increased exports of coal. Within a year of Watt's criticism the FWS was asked to identify ways in which it could change its land management practices to expand the public and economic uses of the refuges and fish hatcheries. The initial responses to this request revealed that expanded use was unlikely without additional resources, but the Service's deputy director found such responses from the agency's regional directors to be unsatisfactory.[32] He sent a second request to the regional directors encouraging them to be more responsive, noting that the FWS was "placing considerable emphasis on increasing economic and public activities" on the lands it manages. In the summary of responses from its directors, the FWS estimated that expanded use could increase the refuge's production of revenue by as much as 35 percent above existing levels. Critics of the anticipated expansion found the plan difficult to reconcile with the department's own findings, namely that environmental problems plague most of the refuges.

Watt's successors may have been less flamboyant, but they were no less concerned about energy and economic development. Donald P. Hodel, Reagan's last secretary of the interior, was a strong advocate of offshore drilling on the Alaska and California coasts notwithstanding the Service's belief that such drilling is a "high-risk operation . . . with potentially devastating impacts on coastal resources." While discussing the Endangered Species Act, Hodel criticized the act as inflexible and complained that decisions of low-level bureaucrats to list species threaten the health and safety of thousands of people. In December 1988, when President George Bush announced that Manuel Lujan, Jr., would be the new secretary of the interior, the reaction of many environmental groups was one of apprehension. During his many years in the House, Lujan consistently received failing marks from these groups because of his opposition to environmental programs.[33] As a spokesperson from one group complained, "We feel we've gotten a lump of coal for Christmas."

Regardless of this claim, one must recognize that interior secretaries are selected not just because of their views on the environment, but for other reasons as well. All such secretaries are political appointees as are all the assistant secretaries within Interior. Partisan preferences extend even to the director of the FWS, who is a political appointee subject to presidential selection and congressional approval.

Political appointees understandably seek to serve their masters, to achieve political goals, and to be responsive to the groups that provided support for their partisan colleague in the White House. These aspirations do not always coincide with the ecological objectives enshrined in the Endangered Species Act, so there is a natural tension between politics and science in the government's efforts to protect endangered species.

Potential Sources of Political Support

The Public

It is axiomatic that public and clientele support are essential to an agency's effectiveness. Nearly all agencies actively curry such support and attempt to gain favorable coverage from the media. Some agencies, because of their missions or responsibilities, frequently find themselves in the news—to their organizational advantage. This advantage can be turned into political capital because high public regard, according to Kenneth Meier, "affects the value other political elites place on the agency and thus its relative position of power."[34] Meier adds that an agency "basking in the warm glow of diffuse public support should be able to press its claims for resources with more authority." With resources comes power, influence, and enhanced opportunities for successful implementation.

The focused attention of specific clientele groups is also important. Ideally, such support comes in the form of a large, well-organized, and geographically dispersed clientele. Desirable groups are those that are cohesive and have a high intensity of commitment to an agency's activities.[35] When an agency can rely on effective group support, the agency will find both its power and autonomy increased and its ability to deal effectively with Congress strengthened.

One of the most effective ways to guarantee and maintain political support is to provide concrete and tangible benefits to recipients who not only recognize the source of their largesse, but who are also willing to inform Congress of their satisfaction. Not surprisingly, agencies that distribute tangible benefits have the most vocal and largest number of clientele groups. In contrast, agencies that impose controls and provide few tangible benefits find it difficult to generate or sustain support. Controls often breed dissatisfaction, especially when the regulatory decisions appear to be arbitrary or are selectively imposed on narrow geographic areas. Affected groups and individuals are likely to be vocal, but in this instance to the agency's detriment. Legislators remain responsive to constituent concerns, so agencies quickly find that opposition to their programs usually provokes congressional consternation.

However necessary public or clientele support may be, not all agencies can create or control it. Such is the fate of the FWS. Although it is engaged in activities related to environmental protection, an issue with high public support, this popularity does not ensure widespread endorsement of the Service's activities. In fact, while strong public support exists for the idea of protecting endangered species as a concept, this support diminishes when the public is asked to reveal its attitudes about particular species. In two national opinion surveys conducted in 1978 and 1980, as an example, 67 and 73 percent of the respondents, respectively, agreed that "an endangered species should be protected even at the expense of commercial activity."[36] Another national survey conducted for the Continental Group a few years later found that 80 percent of the respondents agreed or strongly agreed that: "We must prevent any type of animal from becoming extinct, even if it means sacrificing some things for ourselves."[37]

The problem with the surveys is that they asked a broad question without determining how support might be affected by concern for other values or by the specific species to be protected. These are not insignificant considerations. The Continental Group's survey found strong opposition to extinction, but the survey also revealed that many people do not see a relation between their own activities and the causes of endangerment. During the survey respondents were asked several questions about their preferences for growth and development. These questions and their answers allow the testing of an important hypothesis, namely that those who oppose extinction (or support preservation) will be less likely to support developmental activities than those who are not willing to sacrifice some things to prevent extinction. In other words, as support for preservation increases, support for development should decrease.

The data from the Continental Group survey support this hypothesis, but only in the weakest possible way (see appendix tables A.1 through A.3 for the complete results). Among those who agreed or strongly agreed that species *should* be preserved, even if it means sacrificing some things, these findings stand out:

a) 59.0 percent agreed or strongly agreed that modifying the environment for human use seldom causes serious problems.
b) 60.1 percent agreed or strongly agreed that their communities ought to continue to grow.
c) 54.8 percent agreed or strongly agreed that mankind was created to rule over the rest of nature.

Attitudes about possible limits to growth provide the most unexpected finding, as seen in table 3.2. Compared to the general public and to those unwilling to make sacrifices, Americans willing to make sacrifices to prevent extinctions are *more* rather than less likely to think there are no limits to growth. Clearly, most Americans do not see any direct relation between their activities and the preservation of species.

Additional support for this view comes from a survey that the FWS sponsored. The survey asked a large and representative sample of adult Americans to identify what they consider to be the major causes of endangerment. Most people were unable to identify the chief causes.[38] This limited awareness is not unexpected in view of the public's relative lack of familiarity with endangered species. Of those adults included in the survey, only one-fourth knew that manatees are not insects.

There is further evidence that the public prefers development over preservation. This evidence is found in responses to questions about the trade-offs Americans would be willing to make in regard to the uses of water and endangered species. In the FWS survey, people were asked to indicate whether they would approve of different water uses if they were to endanger a species of fish. Most respondents favored the development of the water over the protection of a species (see appendix table A.4). This is an important finding because many species

Table 3.2 Attitudes toward Preserving Species and Limits to Growth (in percentages)

	We must prevent extinction . . .	
	Agree/ Strongly agree	Disagree/ Strongly disagree
There are no limits to growth for advanced nations like the United States.		
Agree/Strongly agree	57.3	45.0
Disagree/Strongly disagree	42.7	55.0
Total	100.0	100.0
(N)	1,004	240

Source: Computed from data provided by Professor Riley E. Dunlap of Washington State University and collected by Research and Forecasts, Inc., of New York City, in a nationwide telephone survey of adult Americans conducted in 1982 for the Continental Group. See appendix A for further discussion of the survey.

Note: Respondents were asked to agree or disagree with the following statement: "We must prevent any type of animal from becoming extinct, even if it means sacrificing some things for ourselves."

$\chi^2 = 11.78$, $p = .0005$, Somer's d $= .123$. Somer's d provides a measure of statistical association between ordinal or rank-ordered variables. A positive value, as occurs here, indicates that those who agree we should prevent extinction are slightly more likely to agree there are no limits to growth than those who disagree that we should prevent extinction.

of fish are already endangered and because consumptive uses of water also threaten other aquatic and nonaquatic species.

Research on public opinion and endangered species also demonstrates the difficulties in sustaining support for protective activities. To ascertain relative support for different endangered species, the FWS survey presented respondents with the following statement: "A recent law passed to protect endangered species may result in changing some energy development projects at greater cost. As a result, it has been suggested that endangered species protection be limited only to certain animals and plants. Which of the following endangered species would you favor protecting, *even if it resulted in higher costs for an energy development project?*"[39] The nature of the species is an important determinant of the public's response. As species become less "likable," public support for their protection dropped sharply. Ninety-two percent of the respondents favored the protection of bald eagles, and 74 percent favored the protection of American crocodiles. In contrast, majorities opposed the protection of the furbish lousewort and the Kauai wolf spider (see appendix table A.5). Indeed, it may be the case that support for the various species is overstated, at least in terms of the groups or communities likely to be affected. The FWS's sample was a national one, composed of potential beneficiaries of protection. The benefits of preservation to individual members of the public are not likely to be significant, so these individuals have little incentive to organize politically for the Service's advantage. Instead, those who must bear the costs of FWS regulations and decisions to protect species are far more likely to organize and to convey their concerns to key policymakers who, in turn, are likely to expect sympathy for these concerns.[40]

Even when public support for the protection of endangered species is evident, the support tends to be selective. The FWS survey found that support is most likely for species that are aesthetically attractive, that are phylogenetically related to humans, that have economic, cultural, or historical significance, and that are endangered because of direct (as opposed to indirect) harms. Bureaucratic experience parallels these findings. According to Keith Schreiner, a former manager of the FWS endangered species program, emotions run high when endangered animals are concerned: "The warmer the blood, the furrier the hide, the browner the eye, the cuddlier the animal, the higher the emotions run—sometimes almost to fever pitch. Why don't more people care about a highly endangered rattlesnake or a creepy little bug? I'll never understand."[41]

The public opinion data reveal strong support for endangered species in the abstract. Far less certain is the likelihood of firm support for specific actions, such as those requiring the FWS to regulate some activity, to some group's detriment, in order to benefit an endangered species. It may also be that this broad support does not translate into comparable support for the FWS. Most members of the public are surely not aware that the FWS is the federal government's lead agency for the protection of endangered species. Furthermore, the FWS survey found that nearly 40 percent of the respondents had never heard of or had little knowledge of the Endangered Species Act. These are discouraging findings for the FWS, which must depend on or solicit public support to preserve species.

Clientele Support

If public support is ambivalent, is that also the case for support from the Service's most important clients, hunters and fishermen? Ironically, it is this natural constituency that has caused some of the Service's biggest problems in balancing its mandates. No agency likes to find its clientele support to be too narrowly based; the result is often excessive dependence on that clientele and a need to be singularly responsive to it.[42] Nonetheless, several good reasons exist for the Service to be responsive to hunters and fishermen.

Hunters of migratory birds have been required to purchase a migratory bird hunting stamp as a form of a license since 1934. Initially, 90 percent of the receipts (from the $1 stamp) were used to purchase additional wetlands and waterfowl habitats with the remaining 10 percent to be used for managing the refuges. A deal was struck with hunters in 1949—in exchange for doubling the price of the stamp, as much as one-quarter of the acreage of each refuge would be open to waterfowl hunting for the first time. With the precedent set, up to 40 percent of the acreage became available for hunters when the price of the stamp was increased to $3 in 1959. The Service's financial dependence on hunters increased the following year when Congress changed the allocation of funds from the sale of stamps to provide a higher percentage for management services. By the early 1980s, sale of the stamps provided about $15 million per year and more than half of all wildlife refuges were open to fishing or hunting, not only for waterfowl, but also for some big game animals.

The Reagan administration attempted to reassert some control over hunters' access to federal lands, but it was completely unsuccessful. The administration proposed that all hunters and fishermen be required to pay for a federal permit before they could use these lands.[43] Opposition to the proposal was so intense that it was withdrawn within just four days of its delivery to Congress.

For many refuge managers, the political strength of sportsmen affects the distribution of the agency's resources to favor consumptive uses of the refuges' resources. Pressure from local hunters and state game departments has encouraged some managers to transform their refuges into "artificial duck and goose farms." "Digging ditches, building dikes, and planting grain crops," refuge managers, observed Jim Doherty, "engaged in an all-out effort to attract as many waterfowl as possible every year."[44] This perspective appears to have been informally integrated into FWS management plans for refuges. When the GAO conducted an assessment of the Service's management of fish and wildlife, a refuge employee told an investigator that one of his goals was to produce more ducks and birds.[45] Another employee discussed the practice of "shortstopping," that is, attracting and holding waterfowl short of their normal migratory destinations by planting grains for food, by using decoy birds to attract waterfowl, and by preventing water from freezing.

In addition to requiring that duck stamps be purchased, federal laws ensure other close connections between the FWS and hunters and fishermen. The Federal Aid in Wildlife Restoration Act, which became law in 1937, imposes a federal excise tax on ammunition, sporting arms, and archery equipment. A companion act, the Federal Aid in Sport Fish Restoration Act of 1950, imposes a similar tax on sport fishing tackle. Once the taxes are collected, the FWS apportions virtually all the funds among the states for wildlife-related projects. The FWS retains only a small share to cover the costs of administering the program.

Still other indicators point to a potential alliance between the FWS and sportsmen's groups. When Ira N. Gabrielson, the first director of the FWS, retired from the position, he became president of the Wildlife Management Institute, a group supported largely by gun manufacturers with an interest in ensuring the availability of targets for their customers. More recently, after his tenure as assistant secretary for fish and wildlife and parks in the early 1980s, G. Ray Arnett became executive vice president of the National Rifle Association, whose membership includes thousands of hunters. The words of other FWS officials suggest at least a dislike for environmental concerns. Frank Dunkle, who became director of the Service in 1986, came to the FWS with ten years' experience as director of Montana's Fish and Game Department (as well as eighteen months' experience as executive director of the Montana Mining Association). In the latter capacity, he described environmentalists as ecofreaks conspiring to subvert the nation's economy, and he complained that endangered species were halting water projects.[46]

Other interest groups are also potential supporters of the FWS, but not necessarily in the ways that one might expect. The National Wildlife Federation provides an interesting illustration. Although its goals have changed considerably since its creation in 1938, the Federation once functioned as a surrogate advocate for the gun industry.

Through much of the 1940s, the Federation received financial support from gun-related groups. Even after the end of this support, the Federation still "retained its reputation as essentially a group of and for hunters."[47]

The responsiveness of Congress and the FWS to hunters may have satisfied one clientele, but not necessarily another. Environmental groups such as the Sierra Club and the National Audubon Society also represent potential sources of support, but their relations with Interior, and especially with the FWS, are often adversarial. These groups are typically dissatisfied with the government's implementation of environmental programs. Since 1984, for example, the Defenders of Wildlife have issued periodic assessments of the FWS's efforts to protect endangered species. In its initial report, Defenders found fault with nearly every aspect of the Service's endangered species activities. Subsequent reports reveal similar critical conclusions. The Audubon Society's magazine published a lengthy report in 1983 on the Service's management of the National Wildlife Refuge System. The report concluded that many of the refuges are poorly managed, overexploited, inadequately staffed, and largely incapable of performing their primary functions.[48] In making these judgments, the Audubon Society is surely conveying much of the dissatisfaction that characterizes the view of other environmental groups toward the FWS.

The frequent result of this dissatisfaction is lawsuits filed in the belief that the FWS or Interior is not sufficiently vigorous in enforcing legal requirements. The groups' testimony before congressional committees is also generally critical in nature, indicating that the FWS is not conscientiously implementing its obligations to protect endangered species. In addition, some environmental groups tend to be highly selective in their support. Many groups exist to encourage the preservation of birds and mammals, but few specifically devote their attention to insects and crustaceans, species that could benefit considerably from additional interest group support.

Congressional Support

Environmental groups frequently petition Congress to provide Interior and the FWS with increased financial rations, but presidents and Congress are rarely responsive. Richard Fenno's research helps to put congressional reaction to Interior into some perspective. In *The Power of the Purse*, Fenno assessed how thirty-six bureaus and agencies in seven cabinet departments fared in their efforts to secure appropriations from Congress between 1947 and 1962. These agencies included, among others, the Census Bureau, the Soil Conservation Service, the Social Security Administration, and eight units within Interior, including the FWS. Fenno's expectation was that favored agencies would be well treated, that they would occasionally receive more than they asked for and infrequently receive less. Less-favored agencies would be less well received, and their appropriations would reflect their status.

Fenno found that two cabinet departments, Commerce and Interior, consistently faced budget reductions. As he put it, of "the 123 cases in which budget reductions of over 10 per cent were made . . . well over half (67) involved agencies in these two departments."[49] The cuts were particularly noticeable in Interior. Its eight units were among the twelve that consistently suffered the largest reductions from what they had requested from Congress.

More recent data on budgetary support for Interior shows that little has changed. Jeanne Clarke and Daniel McCool's examination of seven agencies with responsibility for natural resources, including the FWS, the NPS, the BOR and the BLM, considered changes in budgets from 1950 to 1980. Among the seven agencies, the FWS received the smallest total increase in annual budget appropriations despite having started at the same budget level as did five of the other agencies. In the authors' words, the FWS and the NPS "demonstrated relatively unremarkable annual budgetary growth."[50]

At least for the FWS, this unremarkable budget growth has not reflected the Service's increased responsibilities, including endangered species; the need to review thousands of environmental impact statements; tens of thousands of permits for projects potentially affecting navigable waters; involvement in the implementation of the Wilderness Act of 1964, the Alaska National Interest Lands Conservation Act of 1980, and an amended Fish and Wildlife Coordination Act, which requires the Service to review all federal projects and permits involving the modification of water bodies. The FWS has faced a similar increase in its land management responsibilities. The agency managed 475 different areas containing slightly less than 18 million acres in 1960. By 1990, the agency had the job of caring for more than 750 areas containing over 90 million acres, a fivefold increase. By most accounts, the FWS is an agency whose resources do not match its statutory obligations.

Given its dramatic increase in responsibilities, it is not unexpected that FWS officials frequently lament their budgets before congressional committees. Pleas for increased support notwithstanding, the Carter administration proposed a 10 percent reduction in the FWS budget for fiscal year 1977 and a loss of seventy-six positions. This is not at all an uncommon pattern. For fiscal year 1980, Interior cut the Service's budget request for resource management programs by $20 million. The Office of Management and Budget (OMB) believed that Interior had been too generous and cut another $20 million before sending the FWS request to Congress. These cuts occurred during the Carter presidency, an administration generally regarded as sympathetic to environmental concerns.

Congressional allocations for endangered species have been somewhat better than the overall pattern for the FWS. The budget for the program has increased, but in only one year between fiscal year 1974, the first year after the passage of the Endangered Species Act, and fiscal year 1990, did Congress appropriate as much money for the endangered species program as it had authorized, as the figures in table 3.3 show. Similarly, only once in this time period did the president's request to Congress exceed the annual authorization. What the figures in the table do show, however, is that Congress is more generous than are presidents. The lack of enthusiastic presidential support was especially evident during Ronald Reagan's first term, when he proposed extensive cuts in funding for the Service's endangered species activities, but only minor reductions in its overall budget. Congress was not completely responsive to the president's proposals, but what it did appropriate in each fiscal year between 1982 and 1984 still left protection activities with less than they had in fiscal year 1981.

Reagan was not the only president to seek reductions in expenditures for endangered species. Soon after George Bush became president in 1989, he declared that "Nature needs our help. Our mission is not just to defend what's left but to take the offense, to improve our environment across the board."[51] To match this rhetoric, the Bush administration proposed that the FWS reduce its expenditures for endangered species by nearly 10 percent.

The adequacy of budgets is more important than their size. Does

Table 3.3   Budget Authorizations, Presidential Requests, and
Congressional Appropriations for the FWS Endangered Species Program,
1974–1991 (in thousands)

Fiscal Year	Statutory Authorization[a]	Presidential Budget Request	Congressional Appropriation
1974	$ 4,000	$ 4,372	$ 4,660
1975	8,000	5,527	5,606
1976	10,000	7,374	7,467
1977	25,000	9,198	9,331
1978	25,000	12,269	12,223
1979	23,000	16,100	15,650
1980	23,000	16,332	16,891
1981	25,000	20,669	21,208
1982	27,000	19,078 Carter	
		15,055[b] Reagan	17,769
1983	27,000	16,550	18,459
1984	27,000	17,361	19,947
1985	27,000	21,242	23,145
1986[c]		22,938	24,787
1987[c]		23,160	24,764
1988	35,000	23,507	26,545
1989	36,500	25,916	28,710
1990	38,000	25,946	32,886
1991	39,500	33,114	—

[a]Grants to the states are not included in the figures for authorizations, presidential budget requests, or congressional appropriations.
[b]The 1982 fiscal year budget requests from the two presidents are not directly comparable. Prior to that fiscal year, money for Common Program Services was included in the presidents' requests. In 1982, these services became a separate budget item.
[c]Due to Congress's delay in reauthorizing the Endangered Species Act, previously approved statutory authorizations did not exist in these years.

Congress provide enough money to enable effective implementation of FWS responsibilities for endangered species? Lynn Greenwalt, who was then director of the FWS, was asked just such a question during a congressional hearing in 1976. He replied with candor unusual for an administration spokesman expected to support the president's budget request: "With our present resources, we are able to carry out about one-fourth of what needs to be done, dealing only with the highest priority species. Carrying out our annual program with our present budget means that complete fulfillment . . . of the act . . . will stretch many years into the future. . . . There will undoubtedly be species lost as a result of delays that might have been saved otherwise."[52] Congress was at least partially responsive to this request for additional support. Authorizations for the FWS's implementation of the Endangered Species Act were increased to $25 million for fiscal year 1977 from the $10 million authorized the previous year. Nonetheless, Congress did not match the increased authorizations with a comparable increase in appropriations, as can be seen in table 3.3.

Since this action Congress has increased substantially both authorizations and appropriations for the protection of endangered species. These increases, however, have not matched increases in programmatic responsibilities or what several administrators believe to be necessary to implement a suitable program, regardless of which political party is in power.

The endangered species activities of the FWS share the same budgetary fate as its parent agency, namely inadequate resources and personnel. Perhaps more important, budgetary resources serve as a measure of executive and legislative support for the protection of endangered species. Budgets provide annual opportunities for politicians to demonstrate such support or, alternatively, to reveal only mild enthusiasm for a program, as seems to be the present case.

Budgets are not the only indication of legislative support for or disenchantment with an administrative agency and its programs. Whereas the appropriations process is a periodic one, congressional oversight can be continual. This oversight can involve far more members of Congress than might be involved with the hearings typical of an appropriations subcommittee, the normal venue for budgetary hearings. At least in the case of the FWS, however, it is unnecessary to go beyond these budget hearings to identify an interesting paradox. Congress consistently provides overwhelming support in its passage of endangered species laws, but the FWS does not always benefit from this support.

With its regulatory functions, the FWS finds it difficult to win influential friends. Conventional wisdom informs administrative agencies that they should avoid decisions that offend or embarrass members of Congress in their districts whenever possible.[53] On several counts the Service has been exceptionally unsuccessful in this regard, as the comments from Representative Virginia Smith (R.–Nebraska) to Lynn Greenwalt reveal:

I have been looking forward to this hearing, Mr. Greenwalt. . . . My anticipation and my concern has been heightened by the growing conviction that your agency may well be the most unpopular agency in the Federal Government. In discussing concerns with other Members of the Congress about Fish and Wildlife Service, my colleagues in general in the House as well as my colleagues on this committee have been unable to elicit one single kind word about your agency.

Do you consider the flood prevention, food producing capacity, and recreation facilities of projects, or do you just think about their impact on wildlife? . . . Why don't you consider these beneficial aspects as well as possible adverse aspects and try to work together with other agencies? [54]

Smith was not alone in her criticisms. Other representatives vented their frustrations as well during the same hearing in 1981, as the comments of Representative John T. Myers (R.–Indiana) indicate:

This committee has had unfortunate experiences . . . with the Department of the Interior, particularly with some bureaus within that department. We have seen the tremendous individual costs that have burdened taxpayers and the delay we have seen in needed projects that through activities of the Fish and Wildlife Service . . . [have] been denied to the public.

You don't come before our subcommittee for appropriations, or things would be a little bit different. . . . It is sad to see that in our country a butterfly or wild plant has more importance in the scheme of things than the public has and that people who pay the taxes have to play a secondary role to wildlife. [55]

Such complaints created a choice for senior Interior officials. They could defend the Service's endangered species activities and demonstrate staunch support for it or they could side with the congressional critics. James Watt chose the latter and joined in the castigation of the FWS. "We find," Watt told the committee members, "that in several instances the Endangered Species Act has been used by the Fish and Wildlife Service people, national and state, to stop small and large water resource development projects because of threats to endangered species. . . . We are going to bring some balance to this." [56]

In view of this tongue-lashing, there was little that Greenwalt could do. He lamely apologized for the lack of congressional support and conceded the Service's poor relations with the BOR, the Bureau of Indian Affairs, the Corps of Engineers, and the Soil Conservation Service. As Greenwalt agreed, he and his colleagues in the FWS are "often viewed as an obstructionist organization because we call attention to the necessity, the desirability of protecting fish and wildlife values."[57]

Although these comments are from one committee, they do seem to reflect a congressional disdain for much of what the FWS does in the regulatory arena. Support for this view comes from Clarke and McCool's study of bureaucratic power. The authors asked their respondents to indicate which of seven resource management agencies have close relations with Congress. The respondents, all of whom were upper-level administrative officials in the agencies, ranked the FWS as having the worst relations with Congress.[58]

Reliance on Scientific and Technical Information

Few agencies flourish without effective public and congressional support. Providing tangible benefits can be one way to maintain this support. Other agencies find that their support rests on the possession of a highly technical body of knowledge vital to the nation's survival or to its economic well-being.[59] Such agencies as the Department of Defense or the Federal Reserve Board are often viewed as indispensable. Their analyses are deemed important, their advice is sought, and their recommendations are taken seriously. In short, technical and scientific units have certain administrative advantages. Such agencies also benefit because of their ability to rely on cause-and-effect relationships, especially when their activities and expertise solve problems or produce economic and technological successes. In a sense, then, many of these agencies are able to provide tangible, albeit indirect, benefits to an appreciative public, most of whom cannot understand the science behind the success. In these instances the agencies have not only the public's support, but its confidence as well. Furthermore, due to the public's alleged lack of comprehension of science and technology, public participation in the agencies' decision making is viewed as either unnecessary or undesirable. The agencies have been able to legitimize many of their decisions by emphasizing their complexity and technical nature.[60]

Science and technology are desirable assets to some agencies, but they do not guarantee administrative autonomy or successful implementation. It is also too simplistic to classify agencies as either non-technical or scientific/technical. Important differences exist among the latter.[61] Some agencies prefer or are required to justify their activities on the basis of scientific or technical analyses but find that opportunities to do so are constrained. With a perceived sense of urgency, as with the desire to find a cure for AIDS, scientific influence is likely to increase. Similarly, when the products of scientific agencies "coincide, promote, or do not interfere with the interests of private economic enterprise,"[62] administrative agencies will have enhanced opportunities to influence public policymaking. Such influence is less likely when policymakers believe that a problem is of less urgency or that they have a "long" time to respond.

Scientific agencies will also be less influential when their programs constrain private enterprise, thus stimulating an active and hostile reaction from vested interests. The federal government's attempts to regulate the use of tobacco offer one of the better illustrations of this point. Compelling evidence links the use of tobacco with the incidence of cancer, yet the government's inroads on tobacco have been minimal, largely due to well-organized resistance. When the basis of action is an identifiable cause-and-effect relationship (as with tobacco and cancer) and an administrative agency is only moderately successful, one can readily imagine how difficult it might be to impose regulatory constraints for the benefits of endangered species on the basis of: a) imprecise diagnoses of a problem; and, b) hypothetical relations that some view as indefensible or that may remain untested for generations. If the resulting regulations also have the potential to jeopardize the activities of popular distributive agencies, then the prospects for success are further imperiled.

Scientific Capabilities of the FWS

The FWS faces many of these problems. As noted earlier, scientists are likely to be influential when they solve problems or enhance economic or technological progress. The FWS often finds itself, however, in just the opposite situation. Every time it declares another species to be endangered, the Service is identifying rather than eliminating a problem that needs immediate attention. Such action can quickly give any agency a reputation for being an organizational pest.

Similarly, the FWS does not fare exceptionally well when Congress distributes resources, with the consequence that the Service's staffing and capabilities are affected. The agency's workload is of such magnitude that it is not able to respond to many requests from state and federal agencies for biological data on how projects will affect fish and wildlife. One assessment found that the FWS was able to provide complete responses to slightly over half of all requests for scientific advice over a three-year period.[63] Although it has had high rates of response to requests for scientific assessments of environmental impact statements and to intraservice requests for assessments of the likely consequences of pesticide use, in other areas response rates are very low. During fiscal years 1978–1980, the Service responded to less than 8 percent of the more than 37,000 requests to evaluate applications for national pollution discharge elimination system permits under the provisions of the Clean Water Act. The Service had similarly low response rates for requests for scientific assessments of ecological emergencies (7.7 percent of requests completed), resource conservation projects (28.2 percent), and miscellaneous permits or licenses (18.6 percent).

In response to criticisms about its low response rates, the FWS noted that it lacks competent people in several scientific fields and has limited capabilities to evaluate the effects of water pollution on fish and habitat.[64] To be responsive to all requests for scientific advice, the FWS estimated that its Division of Ecological Services, which handled many of its consultations, would need more than twice as many employees and a budget nearly 125 percent larger than it had. Despite this estimated need, Congress had never been asked to provide the necessary additional resources.

Even when the FWS is responsive to requests for scientific reviews or assessments, many instances exist in which it does not monitor projects to ensure that its recommendations are biologically sound or have been implemented successfully. Indeed, once the FWS makes recommendations to project agencies, it rarely conducts follow-up reviews, at least if the GAO assessment of the Service's scientific activities is any indication. The GAO found that less than one recommendation in ten is monitored, with the consequence that many recommendations are ignored, neglected, or improperly implemented, to the detriment of fish and wildlife resources.[65] Again, the FWS cited lack of staff and funding as the reason for the inability to monitor its recommendations.

A further explanation of the Service's scientific capabilities lies in its organizational structure. For much of the 1970s, responsibility for FWS research was divided between two units that did not always share compatible objectives or agree on what research should be conducted. One unit focused on regional habitats and fish and wildlife ecosystems while the other emphasized fish and wildlife species and their management. More important than the split jurisdiction is the effectiveness of the research. In this regard the FWS has not received favorable evaluations, either from its own personnel or from those dependent on FWS research.[66] FWS officials once conceded that their research programs were ineffective and poorly coordinated because of overlapping efforts and a lack of goals. State and federal agencies' appraisals of FWS research activities have likewise been critical. People from these agencies complained, according to the GAO, that the FWS is not a focal point for research on fish and wildlife because the agency lacks appropriate expertise, is insensitive to other agencies' objectives, lacks adequate direction, focuses primarily on species rather than habitats, and is not sufficiently appreciative of habitat management needs.

The problems that confront the Service's scientific efforts also affect its ability to justify protection of endangered species. Unfortunately for the FWS, conclusive scientific evidence about vulnerable species is often difficult to obtain. Many endangered species have few individual members, thus reducing the number of research opportunities while increasing the risk should the research adversely affect or even lead to the death of individual members of the species. The California condor provides an example of the latter situation. Ironically, the fewer the remaining members of an endangered species, the greater the need for research and scientific knowledge about the species if it is to be saved.

The quest for this knowledge can be a frustrating one. At best, researchers often find that they must speculate about the answers to such questions as:

1. Which of the millions of existing species are naturally rare?
2. Which species are on the brink of natural extinction and are, therefore, unlikely to derive long-term benefits from efforts to preserve them?
3. What are the causes of endangerment for each affected species, and what must be done to protect them?
4. What is the relation between a species' continued existence and its need for an essential habitat?

5. What is the minimum effective size for this habitat?
6. Can a species threatened with extinction be bred in captivity or be transplanted to another site successfully?
7. What are the consequences of human intrusions on a species' well-being?
8. What level, if any, of development or human intrusion on a species' habitat is acceptable?
9. Which endangered species will benefit from recovery efforts?
10. Which recovery efforts are appropriate, and can they be initiated before a species has dropped below its long-term survival threshold in terms of numbers, genetic diversity, or sexual distribution (e.g., all males)?

The Consequences of Scientific Uncertainty

The frequent inability of researchers to provide data relevant to these questions raises substantial problems for the FWS, while simultaneously suggesting several hypotheses. First, scientists' influence will be diminished to the extent that a lack of consensus about endangered species issues exists among the scientists. There are some ecologists, for example, who remain unconcerned about accelerating rates of extinction. Claiming that extinction is an inevitable and desirable part of evolution, these scientists see nothing wrong with the premature demise of so many species.[67] With such disagreement about the potential implications of anthropogenic extinctions, scientific evidence gives way to other values, primarily political and economic. Laws can require that decisions be based on scientific data, but to the extent that these data are questionable or hypothetical, the law's mandate will be less meaningful.

Second, a lack of compelling evidence undermines the legitimacy of the Service's recommendations. This is likely when policymakers have access to people who claim "expertise" in a given field of policy.[68] The prospect of widespread extinctions alarms many scientists, but some people assert that the protection of biological diversity is not an issue that deserves increased attention. One adherent of this view is Julian Simon, a university economist, who contends that environmentalists mislead the public and oversensationalize the severity of the situation. Simon claims that current levels of extinction are exaggerated and do not justify many of the proposed remedies. To buttress his case, Simon examined scientists' projections of anticipated species' losses and concluded that they are undocumented, largely indefensible, and based primarily on guesswork and conjecture. Simon also concluded that "there is no convincing case for the rapid extinction of species and therefore no need for a solution to what may be a non-existent problem."[69] Even if rapid loss of species were a problem, Simon is not convinced that would justify concern. While not discounting the importance of species, Simon believes that "the extinction of a species is not necessarily a bad thing for humankind and for biological development generally."[70]

Simon's views are obviously controversial among those who believe that species' extinction is a problem of enormous magnitude. Critics say that Simon misinterprets much of the available evidence while ignoring elementary biological principles. However accurate (or inaccurate) his views, Simon's academic credentials and the thrust of his larger message, namely that many alleged environmental ills are vastly overstated, provide support to those interests who oppose efforts to increase protection of endangered species.

Third, scientific uncertainty about the fate of selected species encourages hesitancy rather than initiative among regulators. Scientific uncertainty can be a justification for action, at least for the prudent regulator who believes that it is better to be safe than sorry. Conversely, the same uncertainty can be used to delay action on the grounds that insufficient proof is available to justify action. Organizations like to avoid conflict and uncertainty, so the latter approach typifies the behavior of many regulatory agencies that must base their decisions on hypothetical consequences. Graham Allison states the case well: "[O]rganizations are quite reluctant to base actions on estimates of an uncertain future. Thus choice procedures that emphasize short-run feedback are developed."[71] It is argued that in order to avoid uncertainty, organizations that confront it will negotiate and bargain with other organizational actors to "regularize their reactions" and to obtain their compliance. The need to negotiate with other organizations is a function not only of an agency's power, but also the agency's dependence on other units for successful implementation. Negotiations will be successful when the organization possesses effective inducements or sanctions and when what it is seeking in the way of cooperation does not jeopardize the jurisdictional heartland of other agencies or threaten their goals. In Allison's words, "Projects that demand that existing organizational units depart from their established programs to perform unprogrammed tasks are rarely accomplished in their designed form."[72]

Another organizational incentive for careful consideration exists when agencies must hedge their scientific findings. The FWS experience with the chihuahua chub, a fish found in southern New Mexico, is illustrative. The Service once considered the species to be the rarest fish in the United States, thus deserving the highest levels of protection. Before the species was proposed as endangered, the FWS asked the Federal Disaster Assistance Administration (FDAA) to exercise caution in the implementation of several flood relief projects, lest these projects further threaten the species. Specific recommendations to the FDAA would have been desirable, but the FWS indicated that the chubs' habitat requirements were not well known and that no information was available on their reproduction. "We realize," wrote an FWS employee, that "these data are sketchy but hope they prove useful to you. . . ."[73] Later, despite FWS concern about the chub, a decision was made to list it as threatened rather than as endangered, at least partially because of complaints about the quality of the supporting scientific data.

Prospectively, agencies will also assess the probability that their regulatory actions will generate challenges from interests that may be adversely affected.[74] Agencies do so because organizations typically act to avoid conflict. Weak, qualified, scientific data are, however, an almost certain invitation to challenge or delay government regulations. Challenges can come in the form of legal action or even the threat of legal action, which occasionally is just as effective as successful litigation. Eugene Bardach and Robert Kagan emphasize the consequences of such threats: "Contested cases impose an enormous burden on an agency's understaffed legal section. . . . It may seem more efficient for enforcement officials to concentrate on routine cases that are unlikely to be contested." Routine cases, the authors add, "contribute relatively little toward the achievement of regulatory goals," but such cases do lend themselves to impressive output measures and to a quick disposition of cases.[75] Many cases are not routine, however, and for these an agency's response may well be different. When specific decisions have the potential to be controversial, particularly in critical instances that do not have standard features (e.g., determining that a species is endangered), the administrative response is likely to be handled with great care and over a long period of time. Moreover, the tendency to opt for routine, noncontroversial cases suggests that action is most likely in those situations in which public and political support is high, when potentially well-organized opposition groups are absent, when the regulatory actions will not threaten another agency's mission, and when scientific "proof" is compelling.

There is a corollary consequence of scientific uncertainty, namely that an effective response to uncertainty for many organizations can be found in standard operating procedures. These procedures regularize and simplify organizational action. Such procedures are also highly useful when decisions or problems are routine because there is no need to exercise much, if any, discretion. When problems are not uniform, but rather diverse or even unique, as with endangered species, standard operating procedures become less relevant and the need for discretion increases significantly. Depending on one's perspective, discretion brings with it either sensible flexibility or continuing chaos. Advocates of protection typically claim the latter; regulated groups prefer the former. In either case, discretion is likely to promote unpredictability and still further uncertainty. For the protection of species, this will mean that what constitutes endangerment and appropriate protective measures depends on such capricious elements as budgets, prevailing political and economic winds, personal rather than scientific judgments, and competing interpretations of vague or ambiguous guidelines.

Conclusions

Consideration of several features of the Service's organizational existence—its institutional location, the mission of its sister agencies, its public and legislative support, and its scientific capabilities—all suggest that the FWS and Interior are not well suited to address effectively what many people consider to be the most serious of all environmental problems. Given its potentially undesirable institutional locale, its placement among competing prodevelopment interests, its fragile support, and the hypothetical nature of much of its scientific data, the FWS must possess and exercise effective diplomatic skills if endangered species are to be identified, protected, and eventually restored to adequate numbers. The chapters that follow assess the Service's performance in matching its resources with its responsibilities. Before discussing these responsibilities, however, one additional, intervening chapter is necessary.

With limited staff and financial resources, success will come only if these resources are used effectively. To paraphrase one expert on evaluation research, when resources are insufficient to cover all species in need of assistance, "selection is generally regarded as most efficient if treatment is given mainly to [species] with the highest probability of successful outcome."[76] Such an approach makes the most effective use of an agency's resources and also ensures that those species most in need of attention will receive it. Accordingly, the next chapter briefly focuses on several distinctive ways of establishing priorities for protecting biological diversity.

The institutional environment of the FWS affects the authority of the agency, from designating species as endangered to ensuring their long-term survival. This institutional environment likewise constrains the FWS; without adequate resources and political support, success is improbable, even with conscientious effort. For the FWS, then, the task becomes one of matching its limited political capital with what needs to be done. It is thus necessary to determine an appropriate basis for action. What species should receive attention? At the margins, distinctions are easily made. Cows and houseflies are less deserving of attention than are California condors and black-footed ferrets, both of which are close to extinction.

In contrast to the easy distinctions, many more are fraught with difficulty. Do condors or ferrets deserve greater attention and more resources? Choosing between only two "deserving" species can be problematic, but it is also unrealistic. The fact is that thousands of species are in need of protection, many immediately and others in coming years. Given the need to make responsible choices among thousands of species, are there any reasonable guides for policymaking? There may not be, but no policy initiative can succeed in the absence of specified goals. The discussion in the previous chapter suggests that goals should be established in the context of the agency with responsibility for the program. To do otherwise ignores many relevant considerations. Before doing so, however, it is important to outline what goals might be appropriate from a perspective that is primarily scientific and that allows the nature of the problem to affect the selection of goals rather than allowing such things as politics, economics, or aesthetics to determine which endangered species will be favored. What follows sets forth working principles, based primarily on scientific considerations, that can be used to assess any program designed to preserve biological diversity.

Assessing the environmental consequences of past extinctions and the desirability of biological diversity, advocates of protection and preservation would surely recommend policies that prevent or limit the anthropogenic causes of extinction. Such solutions might include policies designed to halt population growth or to restructure the incentives for economic growth.[1] Restraining population and economic growth would reduce the destructive pressures on habitats. On a smaller scale, because the destruction of habitats jeopardizes many species, public policies should ensure that such destruction does not occur, thereby precluding endangerment. Likewise, when clear, free-flowing streams and rivers are necessary for the survival of certain fishes, then excessive pollution and the construction of dams should be avoided. Such tactics would surely prevent the demise of many species and perhaps even lessen the need for programs designed to save vulnerable species. These recommendations, however desirable they may be from a species' perspective, are infeasible, at least in the United States. Few Americans are willing to lessen their consumptive demands or to eschew an economic system they associate with high standards of living. And, as noted in chapter 3, few Americans realize that their life-styles and demands on environmental resources contribute to species' endangerment. In addition, land developers are in the business of developing land, not in letting economically productive resources lie fallow. Similarly, few people would be willing to risk destructive floods (absent a dam) based on the premise that a species might be saved by foregoing construction of a dam.[2] More important, many species are well beyond the point where prevention can provide a suitable cure.

The solution lies in solving or addressing an existing problem—many species are already in need of intensive care. For these species, prevention is an opportunity foreclosed, but goals and guides to action are still needed. Among all candidates for protection, which should receive it first? The staunchest advocates of protection reject this as an irrelevant question. They contend that all species have inherent value and should be preserved, regardless of cost. To do otherwise, these advocates insist, "is inherently stupid and potentially dangerous."[3]

The protect-all-species position has the appeal of simplicity. In what

David Ehrenfeld has called the Noah principle, named after the first person to have employed it, choices among species are unnecessary. All species are assumed to be equally deserving of protection because they exist. "Long-standing existence in Nature," says Ehrenfeld, is deemed to carry with it the "unimpeachable right to continued existence."[4] Proponents of this view believe that all species are intrinsically valuable, and their known or presumed utility (or disutility) for mankind is unrelated to their rights to exist. This means, of course, that human values are not preeminent, and that a species' continued existence is independent of human judgment about its relative worth. The least attractive and rarely sighted endangered snail is no less nor any more important than the attractive and symbolically appealing bald eagle.

A related reason for the protect-all-species approach focuses on the interdependence of all parts of ecological communities. Although scientists will never fully understand the complex linkages that exist among species, or even the ecological importance of many identified linkages, many scientists are comfortable in declaring that "everything in nature is connected to everything else." If a species exists it has some ecological function and some reason for being. Eradication of a species may have unforeseen consequences on other species or on ecological stability. However despicable a species may be—locusts, smallpox virus, and *Anopheles* mosquitoes come to mind—eradicating it is undesirable because doing so might "entail some small chance that the inevitable consequent changes in Earth's ecosystems would make the world less hospitable for humanity, causing worse suffering than that previously inflicted" by the species.[5]

The save-all-species approach may be ecologically astute, but it has several pitfalls. The approach most obviously ignores the shortages of essential resources and the constraints this imposes on recovery programs. The approach also ignores public opinion, the strong support that exists for the protection of some of the more appealing species and, in contrast, the difficulty in sustaining support for the majority of other, less endearing creatures. A further problem is that not all endangered species share the same vulnerability to extinction or the same level of depletion. Clearly, other approaches must be considered.

A second approach to protection asserts that decisions to classify species as deserving of protection should be based on the best available scientific data, particularly scientific research suggesting that premature extinction will occur unless preventive actions are taken.

Scientific information about a species' plight legitimates decisions to preserve biological diversity. Reliance on scientific data as a decision criterion is widespread in the United States. Environmental laws commonly require agencies to justify their regulatory standards on the basis of such data. The Occupational Safety and Health Act mandates that standards dealing with toxic materials assure "to the extent feasible, on the basis of the best available evidence, that no employee will suffer material impairment of health or functional capacity. . . . In addition . . . other considerations shall be the latest scientific data in the field." Likewise, the Clean Air Act Amendments of 1977 dictate that air quality criteria "shall accurately reflect the latest scientific knowledge. . . ."[6]

Scientific data are usually incomplete, so environmental laws often include a requirement that a margin of safety be incorporated into regulatory standards. Such a requirement reflects a preference for prudence. For the protection of species, such a rule would mean that, in the face of scientific uncertainty about which species are endangered and uncertainty about the consequences of extinction, species should be given the benefit of doubt.

A further working principle related to protection emphasizes the actions to be taken once a decision is made to protect a species. Simply stated, remedial measures ought to be commensurate with need and sufficient to lead, when possible, to full recovery of a species to the point where it is no longer on the brink of extinction. The logic of this principle should be evident—halfhearted or perfunctory measures can delay a species' extinction but are unlikely to prevent it. Moreover, if unsuccessful, such measures provide no effective return on a government's investment of time or resources.

Even if one accepts these few working principles, at least two kinds of problems exist with the principles. Asserting that all species deserve protection because they exist or that scientific analysis should be used to justify protection ignores the fact that there are more species in danger than scientists can examine or funds available to protect. In addition, reliance on ethical or scientific justifications implicitly assumes that agencies with responsibility for endangered species act rationally, pursue only one set of goals, and are oblivious to the political environment in which they operate. These are unrealistic expectations, as the findings in the previous chapter reveal. Both categories of problems also imply a need to consider other approaches to saving species, namely approaches that recognize a likely shortage of resources and a realistic assessment of what administrative agencies can reasonably do.

Priority Systems Related to Scientific Criteria

With a likely shortage of financial and professional assets and, as some believe, an inability to save all species, there is a need to create some system that effectively uses available resources. To do otherwise randomly condemns helpless creatures to extinction regardless of their relative merits, potential contribution to humans' well-being, or contribution to ecological stability. The other possibility is to develop a priority system that assists policymakers in deciding which species are more important and more worthwhile to save or, alternatively, which endangered species are less important and, therefore, deserve less attention than others. Norman Myers states the problem well: "Since we are clearly going to lose many hundreds of thousands of species before the end of the century, we need to know which ones we can 'best afford' to lose, which ones would certainly leave major ecosystems with critical injury if they disappeared, which ones should be saved because their loss could precipitate ecological breakdown whose dimensions we can hardly start to envisage, and which ones should be preserved virtually at any cost."[7]

Based largely on ecological considerations and the anticipated consequences of extinction of different kinds of species, scientists have proposed a number of ranking schemes. One such scheme maximizes the preservation of genetic diversity.[8] In this scheme, monotypic genera (i.e., a genus with a single species) receive priority over species belonging to genera comprising several species, which in turn would receive priority over subspecies. This means that when choices must be made, it is more desirable to save a single, highly distinctive species such as the California condor or the moapa dace rather than a species that has several (or many) close genetic relatives.

Genetic diversity is crucial to the processes of evolution and the ability of species to survive changes in their environment or to adapt to local conditions like rainfall, predation, temperature, or competition. The greater the genetic diversity, the better the prospects that species will withstand environmental change. This relationship has an important corollary for human well-being. Genetic variability provides opportunities to develop improved or pest-resistant strains of existing crops. As an example, some strains of U.S. wheat remain resistant to rusts or fungi for about five years, at which time it becomes necessary to develop new resistant strains. Without the introduction of new genetic types this would be impossible.[9] Genetic diversity becomes especially important when farmers depend on relatively few plant species, as is now the case. Such narrow monocultures create a high vulnerability to pests, diseases, and changes in growing conditions.[10] Ironically, scientists' ability to respond to these threats is being reduced through the extinction of many wild strains of rice, wheat, yams, and fruits.

A scheme related to genetic diversity favors K-selected species over r-selected species. The latter "are usually short-lived, have brief gaps between generations, and feature high rates of increase." For these reasons, r-selected species are less prone to extinction and are rarely endangered. As Myers indicates, such r-selected species as rats, rabbits, and cockroaches, all of which are small in size, successfully confront environmental disruptions "because of their opportunistic attributes and their built-in capacity to expand their numbers rapidly. . . ."[11] Should any of these species face reduced numbers, they can quickly rebound under favorable conditions.

K-selected species, in contrast, are especially prone to extinction because of the common features they share—long life spans, relatively low reproductive potential, and considerable time between generations. Whales, condors, and whooping cranes, all of which are K-selected species, "make unusually efficient use of particular environments" and, as Myers observes, "this means they are closely adjusted to the long-term capacity of their habitats to support them." Not only are K-selected species relatively few in individual numbers, but because of their characteristics, when one of these species loses "a large proportion of its numbers, it may prove critically unable to build up its stocks again, no matter what protection measures are provided."[12]

Still another ranking method asserts that species' relative functional contribution to ecological stability should be the primary criterion for allocating resources. Not all species make the same contribution —it depends on their numbers, biomass, status in the food pyramid, and so on.[13] Some species are especially important because of their role as keystone or indicator species. The former are key links in ecological interdependence; their elimination is likely to be followed by the extinction of many more species whose existence or well-being depended on the keystone species. The sea otter is often cited as an example of a keystone species.[14]

Indicator species can alert humans to potential problems in the environment in the same way that canaries were once used to warn miners about the presence of harmful gases. In the past, the brown pelican, the peregrine falcon, and other birds of prey have alerted humans to the dangers associated with DDT and other toxic substances.

The problem with these approaches is that it is often difficult to identify keystone or indicator species. So little is known about species' interdependence in many ecosystems that keystone species cannot be identified unless their numbers are accidentally (or intentionally) reduced, at which time it may be too late to save the species from extinction. Indicator species can alert people to ecological harms, but awareness of a problem does not ensure an understanding of its causes or possible solutions. Furthermore, indicator species tend to be at the top of food chains since these are the species that concentrate contaminant materials in their tissues.[15]

Favoring species at the top of food chains contradicts a priority scheme based on trophic levels that gives preference to species at the base of food chains and low on the phylogenetic scale.[16] This scheme favors plants, the foundation of all food webs. Green plants serve as producers, transform the sun's energy through photosynthesis, and directly support herbivores (animals that eat plants) and, indirectly, carnivores (animals that eat other animals). Transfer of energy from lower to higher trophic levels is highly inefficient, so about one hundred tons of producers are required to support ten tons of herbivores and about one ton of carnivores. Loss of producers inevitably leads to reduction in the numbers of herbivores and carnivores and, in some instances, a guarantee of extinction for herbivores that are exclusively dependent on selected plant species. Accordingly, such a view favors plants at the expense of carnivores and predators. Peter Raven, a leading botanist, succintly summarizes the argument for this position: "the diversity of plants is the underlying factor controlling the diversity of other organisms and thus the stability of the world ecosystem. On these grounds alone, the conservation of the plant world is ultimately a matter of survival for the human race."[17]

Other scientists agree with Raven. They believe that saving species at the top of the food chain is not as important as saving those at the bottom. As the Conservation Foundation once noted, "the loss of a species at the top of the food chain—like the condor, the grizzly bear, and the bald eagle—is not likely to affect the underlying ecosystem or man."[18] Such an opinion, with which many scientists would disagree, provides an interesting example of the differences in scientific opinion, namely whether those at the top or bottom of food chains ought to be given preference in protective schemes.

A further ranking scheme focuses not on trophic levels, but rather on species' propensity for extinction or the degree and immediacy of threat to their continued existence. Not all species are equally vulnerable to extinction. Some species, because of specialized habitats, natural rarity, genetic characteristics, restricted distribution (such as on islands), or sensitivity to environmental disruption, are especially vulnerable to extinction.[19] Many species illustrate these characteristics. The Hay's spring amphipod, a crustacean, is found only in a single spring in Washington, D.C.'s National Zoological Park. The Florida Everglades kite, a bird, feeds on a single species of snail. Populations of whooping cranes have probably never exceeded 1,500 individuals.

These and most other similarly vulnerable species are likely to avoid premature extinction in the absence of environmental disruptions. But such disruptions as urbanization, alteration of habitats, channelization of rivers, and use of off-road vehicles constitute threats to these species' existence and are likely to increase their vulnerability to extinction, particularly for species with a limited geographic range. Many people thus believe that the degree and immediacy of threat should determine which species become the beneficiaries of recovery efforts. Going one step further, several scientists have suggested a triage system, much like that used in World War I to decide which wounded soldiers would receive medical care. Norman Myers, a strong advocate of protection, is among these, and he recommends that species threatened with extinction be put into categories—those that protective and recovery efforts are likely to benefit and those that are beyond help. Endangered species placed in the first category would receive the benefit of whatever resources are available to save them; those species in the latter category would "not be admitted to the hospital," and would consciously be allowed to become extinct on the presumption that no assistance would be beneficial. "Insofar as man is certainly consigning huge numbers to oblivion," says Myers, "he might as well do it with some selective discretion."[20]

Making judgments about the proper categorization is both problematic and controversial. Scientists recognize that when populations of many species are reduced below certain levels their ecological significance is diminished or eliminated and their continued breeding success, genetic variability, and adaptive fitness are threatened.[21] All these factors might suggest the appropriateness of putting such depleted species in the latter category. At the least, doing so would be anathema to those who believe that no species should be denied protective measures. Responding to such a view, Erik Eckholm observes that although it may be galling "to anyone steeped in the mysteries of biology, consciously writing off some life forms [through triage] in order to save many more may be the best among unpleasant alternatives."[22] Despite his choice of priority schemes, Eckholm realizes that triage is no more than a reaction to, not a cure for, the causes of extinction.

Another ranking system based largely on scientific criteria asserts that a focus on individual species is too narrow. Instead, the system would attempt to preserve important habitats and ecological communities. The logic behind this scheme is that species are dependent on their habitats, share interdependence with other species, and make contributions to ecological stability only in the context of their habitats. Put in other terms, habitats rather than individual species have biological value and thus deserve protection.[23] Protecting individual species may not save them but preserving their habitat has a far better chance of doing so, especially when the species are dependent on keystone species yet to be identified. Similarly, protection of habitats, particularly biologically unique ones, is more likely to be advantageous to many species than is a species-by-species approach. Of course, the protection of habitats focuses on the attributes of an ecosystem rather than on species within ecosystems. This is an important consideration when the required habitats are likely to be large, as is the case with many carnivores, and in situations in which the habitats overlap with disruptive human behavior. It may be easier, for example, to justify recovery measures for wolves or grizzly bears than it is to justify protective measures for the thousands of square miles these species need to exist.

All these ranking schemes are based on ecological considerations. As noted, several of them have problems associated with their application. For most, a lack of information hinders efforts to apply the ranking schemes. As an illustration, few keystone or indicator species have been identified, and scientists are often unable to estimate the ecological consequences of an individual extinction. Indeed, there tends to be redundancy in the performance of many ecological functions, so the loss of a species may have no readily identifiable adverse consequences for an ecosystem's viability. Without a commonly accepted definition of diversity, efforts to preserve it will be difficult. From some perspectives, protecting diversity may allocate scarce resources to species such as whales that are "evolutionary dead ends" and relatively incapable of responding to environmental changes.[24]

Reliance on multiple scientific schemes may be desirable, but the schemes are not entirely compatible. Preserving on the basis of genetic diversity can favor species that make only a negligible contribution to ecological stability. Favoring K-selected species can ignore the greater opportunities for speciation associated with r-selected species. Plants are at the base of trophic levels, but species at the top may be more vulnerable to extinction.

A more fundamental problem focuses on the definition of a species. Species are commonly defined as those individuals who resemble one another, who share a common gene pool, and who are able to interbreed among themselves but not with members of other species. In reality, however, some species do interbreed, and what distinguishes one species from another (or a species from a subspecies) can depend on personal predilections rather than objective criteria.[25] The classification criteria available include descent, behavior, morphology, gene pool, and reproductive isolation, but their use frequently depends on who is making the taxonomic classification. This is not an insignificant consideration when protective measures are justified on the basis of existence, or for species rather than subspecies, or for mutants.

Priority Systems in a Political Context

An equally fundamental problem in establishing priority systems based primarily on ecological considerations is that decisions to protect species are made in a politicized environment. It may be desirable to rely primarily on scientific criteria, but only the most naive will expect this to occur. Indeed, one can speculate that as scientific certainty decreases, policymakers' reliance on ecological criteria will not only diminish but also give way to what they consider to be far more relevant considerations, including ones that mix scientific justifications with economic and political values.

Reliance on these values might provide for different ranking systems. One such scheme, admittedly extreme, is solely utilitarian and asserts that only species that are of demonstrated value to humans should be preserved and protected. This approach coincides with the Judeo-Christian belief that humankind's role is to dominate and exploit nature and with the belief of most Americans that plants and animals exist primarily to be used by humans.[26] This seeming arrogance, as John Passmore graphically explains, "makes men think of nature as a 'captive to be raped' rather than as a 'partner to be cherished.'"[27] If a species with no discernible instrumental value becomes endangered and then extinct, so be it. Humans, so the argument goes, are no worse off than before the extinction because nothing of value has been lost. A variant of this ranking scheme acknowledges that species have value, but that the value is a relative one. Given a choice between preserving a species and achieving another goal, the advocates of the utilitarian approach would compare the project's anticipated value to that of the species. Like an economist employing cost-benefit analysis, the utilitarian advocate selects the most valuable alternative, thereby using resources efficiently and simultaneously enhancing economic growth.

Concern for economic efficiency in decision making is not a new phenomenon, but it is one that has received significantly increased attention. Presidents Gerald Ford and Jimmy Carter instructed federal regulatory agencies to be more cognizant of the costs and benefits of their decisions. The Reagan administration went a step further and required that the costs and benefits of proposed regulations be considered during their development. Inherent in all efforts to rationalize decision making is the belief that limited resources must be used as effectively as possible. To do otherwise uses resources inefficiently and undermines the government's credibility.

The utilitarian approach raises at least several concerns. From a scientist's perspective the approach ignores completely species' ecological contributions unless they are known and can be valued in monetary terms. At the least, this is highly problematic since it depends on one's willingness to consider species as resources. Natural resources are assets or commodities, such as gold, coal, or petroleum, that can be converted into money. These resources must be exploited and developed before conversion. When species are considered to be resources, they have value only when they can be exploited. A conventional economic assumption is that unexploited resources are sterile. Failure to exploit resources supposedly does a disservice to future generations because extracted resources can be converted to productive capital and used to provide higher standards of living than present generations now enjoy.[28] The Department of the Interior once accepted such reasoning when it justified preservation of whales on the grounds that their restoration could lead to their return "to a productive capacity and a resumption of sporting or commercial uses."[29] With such a perspective, species without actual economic value become, by economists' definition, nonresources.

To protect species within this framework it is often necessary to contrive economic value for many of them. The difficulty in assigning value is that all values will be based on human interpretations of what is important or worthwhile. In other words, species have no value except as they are useful to humans.

This consideration has not deterred economists from assessing the relative monetary value of some endangered species. J. R. Stoll and L. A. Johnson, as an illustration, assessed the public's willingness to pay to preserve whooping cranes and calculated the estimated national value for the species to be between $573 million and almost $1.6 billion. Two other economists concluded that preserving bald eagles in Wisconsin produce benefits totaling about $28 million.[30] Whether one judges these values to be large or small, the act of valuing them economically may actually increase the threat to certain species. David Ehrenfeld accounts for this situation by noting that "any competing use with a higher value, no matter how slight the differential, would be entitled to priority" in the use of the species, its habitat, or its larger ecosystem.[31] From the perspective of economic efficiency, "higher value" takes precedence, and one mathematician's research reveals what the result can be. Colin Clarke examined whether it makes economic sense to preserve certain species that are known to have some economic value. Under some circumstances, Clarke concluded, it is economically desirable to exterminate an entire species because if "the annual discount rate exceeds the natural growth rate of the resource, the resource then becomes an 'inferior' investment, and its owner (if any) would prefer to 'cash it in,'" that is, to consume the resource and spend the profits elsewhere.[32] Natural growth rates are likely to be below discount rates for species that reproduce themselves only over an extended period of time. Whales and some other marine mammals are examples. Why preserve these species, the economist might ask, if it is more profitable to exterminate them now rather than wait until some later date?

The incentive to exploit will be especially intense when species are viewed as common property resources, available to whomever wishes to exploit them. If, in the interests of preservation, someone decides not to exploit a species, there may be no way to prevent someone else from doing so. Whales again provide the example. When conservation groups urged the United States to impose a moratorium on commercial whaling, American whalers argued that such a ban would be meaningless unless Japan and the Soviet Union did the same.

A further problem with the strict utilitarian approach (and in deciding what species to protect) is that it presupposes someone's ability to ascertain the *future* value of species.[33] Many scientists believe that thousands of species have untapped potential to contribute to an improved quality of life. Admirable though it may be, the scientists' defense is not without weaknesses. Millions of species have never been scientifically described, and generations of further research are necessary to determine whether species already described will contribute to humankind's direct economic benefit. Relatively few species are likely to do so. As David Ehrenfeld observes, people in capitalist societies "are not ready to call something a resource because of long-term considerations or statistical probabilities that it might." Ehrenfeld adds that without actual or potential demonstrated economic value, species "are not easily protected in societies that have a strongly exploitative relationship with Nature."[34]

Another economic approach to establishing priorities for protection asserts that cost effectiveness should be the evaluative criterion. Facing limited resources, an astute FWS should use these resources in a way that maximizes the number of species that can be protected, or so it is argued. To implement this approach, it is necessary to estimate the costs of listing and protecting each species facing premature extinction. Costs could include expenses for scientific research to determine the status of a species, for land acquisition to preserve its habitat, for the implementation of protective measures, for artificial propagation, or for whatever might be necessary to preserve the species. Some species could be protected at low cost, perhaps a few thousand dollars, while other species would require millions of dollars a year for twenty or more years.

Once the costs for each endangered species are estimated, the costs would then be ranked from the least costly to the most expensive. Species would be protected up to the level of resources available, thus maximizing the return on the investment. This approach might appeal to economists, but ecologists could correctly claim that it ignores completely all attributes of a species except the estimated cost of its protection. Biological considerations would be irrelevant, so a monotypic keystone species could find its road to extinction hastened because of the high cost of preservation.

All the schemes so far discussed have the potential to cause problems for the FWS. Approaches to priority systems that rely primarily on biological considerations ignore the institutional environment of the FWS as well as the agency's limited budget. Reliance on economic considerations may be efficient and politically astute, but it is scientifically indefensible. It should be remembered, however, that not all the approaches are mutually exclusive, and it may be possible to combine the desirable features of several approaches. From the vantage point of the FWS the most appealing approach might be one that allows the agency to choose a strategy that not only satisfies its organizational goals but that is also responsive to its political environment.[35] To do otherwise may be to risk losing opportunities for success and public and legislative support. The chapters that follow assess these possibilities.

5 The Listing of Species, Organizational Choice, and Decision Rules

The most crucial stage in the protection of endangered species is the first—the stage at which decisions are made about which species are on the brink of extinction. In the listing process, decisions must be made about which species deserve protection and will benefit from the available resources and recovery strategies. Choices about listing a species initiate all subsequent steps and, if protective measures are suitable, determine which species might be saved from premature extinction.

This and the next chapter focus on the nature of these choices, raising questions about such issues as procedural requirements for listing, the consequences of listing, and the selection and use of various priority schemes. For these and other issues, an important consideration involves the rules that apply to the decisionmaking processes. These rules can either hinder or facilitate the prospects for successful implementation. Decision rules can be biased in favor of program objectives or, alternatively, can be written in such a way as to frustrate the most competent and conscientious program administrators.

In several ways decision rules serve as indicators of executive and legislative support for administrative agencies. Such rules typically reflect political and economic values. In addition, agencies can be given broad grants of discretion and much flexibility, thereby revealing politicians' trust and a belief that discretion is necessary and will be used wisely. Alternatively, administrative discretion can be circumscribed as politicians demonstrate not only a concern for how an agency implements its responsibilities, but also for the outcomes of its activities. It is possible, for example, to impose a constraint on officials by requiring that their decisions be based on a thorough analysis of the likely consequences of the decisions, as is now the case for federal agencies that must prepare environmental impact statements. Similarly, constraints can be added by imposing stringent burden-of-proof requirements, by mandating the kinds of information that must be provided to support an action, by limiting the time agencies have to issue rules or regulations, by imposing elaborate mechanisms for public participation, by restricting the kinds of action an agency can select, by increasing the number of officials and/or other agencies that must consent before a rule can be issued, and by creating opportunities to challenge agency decisions before courts or legislatures. In each instance, whatever the decision rules are, they affect discretion, introduce social and economic considerations into scientific deliberations, shape administrative flexibility and, prospectively, communicate politicians' relative confidence in an agency's judgments.

One can also hypothesize that as the number of constraints increases, the likelihood will decrease that biological considerations will be preeminent in decisions to list species. Likewise, administrators are likely to have a relatively free hand (with favorable decision rules) when the consequences of their actions are relatively insignificant. As the consequences of classifying species as endangered becomes more burdensome to the public and other administrative agencies, however, there is likely to be a decrease in administrative discretion (and a corresponding increase in the involvement of political appointees) due to complaints of agency arbitrariness and insensitivity to other concerns.

Determining What is Endangered

Initial Decisions, 1964–1973

Although the United States did not have a law specifically devoted to the protection of endangered species until 1966, efforts to develop a list of such species had begun several years earlier. Interior's BSFW had established the Committee on Rare and Endangered Wildlife Species (CREWS) in 1964. Two years later, in July 1966, CREWS issued a formal report that listed 83 native species (17 mammals, 36 birds, 24 fishes, 3 amphibians, and 3 reptiles) as endangered.[1] Among these species were such well-known ones as the black-footed ferret, the California condor, the whooping crane, and the blue whale, as well as many lesser known species like the Kauai 'O'o and the blunt-nosed leopard lizard. Additional species were listed as rare and, though not facing the threat of immediate extinction, were so few in number that careful monitoring of their condition was deemed appropriate. No plants, insects, snails, clams, or crustaceans were among those considered to be either rare or endangered. The emphasis was solely on vertebrate species.

The committee noted that its conclusions were reached without regard to BSFW policies or management and administrative restrictions. To illustrate this point, the report listed the Utah prairie dog as a rare species and indicated the reasons for its decline included poisoning and animal control operations. Ironically, it was the animal damage control unit of the BSFW that was funding some of these operations.[2] The committee also admitted that there had been both a shortage of data and disagreement about which species to include. One fish species, the humpback chub, was identified as endangered even though its fecundity, estimated numbers, reasons for decline, former distribution, and potential for captive breeding were all unknown. The committee declared other species to be endangered although it was not sure that some of them still existed. Many other species were believed to be "on the verge of total extinction."

For the species it had identified as endangered, CREWS believed that their prospects for survival and reproduction were in "immediate jeopardy" due to "loss of habitat or change in habitat, overexploitation, predation, competition, [or] disease." Without help for the endangered species, CREWS noted, extinction would probably follow. Despite this warning, the CREWS did not recommend any specific responses other than "united appropriate action" and a "long-range habitat preservation program . . . for species that would otherwise disappear." The committee made this recommendation despite its awareness that the type and amount of habitat required for protection of most species was unknown.

When the CREWS published its list in mid-1966, it had no statutory authority to do so. This deficiency was remedied later that year with the passage of the Endangered Species Preservation Act. The act instructed the secretary of the interior to publish, after consultation with the affected states, a list of native fish and wildlife that he regarded as threatened with extinction. For a species to be included on the list, its existence would have to be "endangered because its habitat is threatened with destruction, drastic modification, or severe curtailment, or because of overexploitation, disease, predation, or because of other factors, and that its survival requires assistance."[3]

The law also gave the secretary the authority to seek the advice and recommendations of interested persons and organizations on what should be listed. The law did not require, however, that the secretary solicit public comment on any proposed listings. There was also no requirement that the secretary had to be responsive to the advice received or to justify his choices through the use of scientific data. One possible explanation for the latitude given to Interior relates to the limited protection the 1966 law provided to species listed as endangered. This law prohibited the taking or possession of an endangered species but only when it was within any area of the National Wildlife Refuge System.[4] To take meant "to pursue, hunt, shoot, capture, collect, kill, or attempt to pursue, hunt, shoot, capture, collect, or kill."

For endangered species not within a refuge, the rules were completely different. Deferring to traditional state jurisdiction over resident wildlife, the 1966 act prohibited the federal government from regulating the hunting or fishing of endangered species on lands outside the national refuge system. Moreover, the law required that, "to the extent practicable," regulations for hunting and fishing within the federal system should be consistent with relevant state laws. These provisions meant that a member of an endangered species might be protected only so long as it remained within a refuge. An endangered animal, such as a wolf or an alligator, moving a few hundred feet across a refuge boundary would be fair game for a state-licensed hunter since it was legal to hunt these species in many states. Even for species on federal land, however, there were doubts about federal authority to protect them in the absence of any threatened harm to the land.[5] Clearly, the initial federal endangered species legislation was directed primarily at listing rather than protecting native species.

There was, however, one exception to the emphasis on listing. Using funds from the recently passed Land and Water Conservation Fund Act of 1965 (LWCF), the Endangered Species Preservation Act authorized Interior to spend up to $5 million a year, for three years, to acquire "lands, waters, or interests therein" that would be needed to conserve or protect species threatened with extinction.[6] Thus, the secretary could use these funds to expand existing refuges or to create new ones specifically for endangered species.

The emphasis on listing rather than protecting species was re-inforced in 1967 when Secretary of the Interior Stewart Udall issued the first official list of native endangered species.[7] A comparison of this list with the 1966 CREWS list reveals only a few changes. Both lists contained the same thirty-six species of endangered birds and the same six species of reptiles and amphibians. The 1967 list also included twenty-two of twenty-four species of fish and thirteen of seventeen species of mammals from the earlier list. Among the mammals, four species of whales were not included in 1967, perhaps because of uncertainty about whether they qualified as native species.

Additional lists of endangered species were issued in 1968 and 1969. The CREWS was responsible for the one issued in December 1968, and this list was also a virtual replication of the two earlier ones.[8] Again, only vertebrate species were listed. Some species, mostly birds, had their status changed from rare to endangered because of the availability of new information. For the same reason, some species were added because they had been rediscovered in small numbers while others were removed from the list because they were better off than originally believed. Four species of whales, the blue, the humpback, the Atlantic Right, and the Pacific Right, were once again identified as endangered.

The 1968 list presumably reflected Interior's official belief about which species were endangered and, therefore, deserving of protection on federal refuge lands. Nonetheless, the revised list did not mandate any changes in hunting or fishing on federal refuges because the Interior Department chose not to publish its 1968 revisions in the *Federal Register*, as the 1966 law required. As a result, some species believed to be on the brink of extinction could legally be hunted or taken anywhere, including within federal refuges, simply because the species had not been officially designated as endangered.

Still another list was issued in March 1969, but this one was pub-lished in the *Federal Register*, thereby making it an official list. However official the list, it was not without ambiguity. Once again, the four species of whales were omitted, and grizzly bears, which had been classified as endangered in 1965, 1966, and 1967, and rare in 1968, were neither rare nor endangered on the 1969 list.

With the passage of the Endangered Species Conservation Act in late 1969, Congress changed slightly the ground rules for listing. The new law provided the secretary of the interior with continued discre-tion. The secretary would be allowed to designate as endangered any species or subspecies of fish or wildlife threatened with worldwide extinction whenever he determined, "based on the best scientific and commercial data available to him . . . that the continued existence of such species or subspecies . . . is . . . endangered," due to any one of several causes, including such reasons as destruction or modification of habitats, overutilization for hunting or commercial reasons, disease, or other factors.[9] The new law also broadened the definition of the species that could be listed by specifically identifying mollusks and crustaceans in addition to the vertebrate species that had been listed in previous years. Whatever the choices made, publication in the *Federal Register* would still be necessary.

Two changes in the 1969 law were especially significant. First, only species or subspecies threatened with *worldwide extinction* could be listed as endangered. A separate population of a species could not be listed unless the entire population was endangered. Grizzly bears might be faced with extinction in the lower forty-eight states, but the species, according to the 1969 law, could not be listed as endangered if it was doing well elsewhere, such as in Canada or Alaska. This choice ignored a salient consideration, namely that separate populations of the same species often have some distinctive biological features and even a slightly different genetic composition.

Second, the law prohibited both interstate and foreign commerce in illegally taken endangered fish and wildlife, including mollusks, reptiles, amphibians, and crustaceans. The prohibitions on commerce in endangered species were not absolute. Responding to pressures from scientific and commercial interests, the law allowed Interior to authorize the importation of endangered species "for zoological, educational, and scientific purposes, and for the propagation of such [species] in captivity for preservation purposes." An important exception was likewise included for people who might suffer "undue economic hardship" because of the law's implementation. For people who had entered into a contract prior to a species' listing as endangered, the secretary could allow importation of the (dead) species for up to a year in such quantities as he deemed appropriate. This last provision was included largely to satisfy the fur industry, which successfully argued that a complete ban on imports would harm its workers and be meaningless if other nations failed to impose similar restrictions.[10]

One change of somewhat less significance involved opportunities for public participation in the listing process. The 1969 law required

Table 5.1 Changes in the Status of Selected Species, 1966–1970

Species	Year				
	1966	1967	1968	1969	1970
Grizzly bear	E	E	E	R	NL
Utah prairie dog	R	NL	E	E	Proposed as E
Caribbean monk seal	E	E	E	E	NL
Little Colorado spinedace	E	E	R	NL	NL
Tule white-fronted goose	E	E	E	E	NL
Desert dace	E	E	E	E	NL

Source: Compiled by the author from various FWS lists.
Key: E = Endangered, R = Rare, NL = Not listed

Interior to propose the listing of species before taking final action, thus allowing time for public comment. Although opportunities for public comment were provided, other means of public participation were limited. Interest groups or members of the public could petition the department's secretary to delist a species by providing "substantial evidence" to warrant such an action. Comparable evidence could not be used to request that a species be listed as endangered. These provisions were entirely acceptable to Interior, which asserted that decisions to list or delist should be left completely to the secretary's discretion.[11]

With its new legislative authority, Interior henceforth modified its listing procedures because of the law's potential impacts. To comply with the law, the secretary sought public comment before finalizing the proposed addition of twenty-two native species to the 1970 list. Most of the species listed in previous years were retained on the new consolidated list. Nonetheless, several interesting changes did occur from the first CREWS list in 1966 to the secretary's list in late 1970, as seen in table 5.1. The Utah prairie dog had been formally listed as endangered in 1969, but it was reproposed for listing in August 1970. After the opportunity for public comment, the proposal was withdrawn. The desert dace, a monotypic, relict fish found in only one county in Nevada had been considered as endangered from 1966 through 1969. Without any explanation, the dace was omitted from the 1970 list, only to be reproposed for listing in mid-1985 because of continuing threats to its habitat.

After declaring a few more native species as endangered in late

1970, Interior's listing activities virtually ceased for the next two years. Between December 1970 and June 1973, no native species were listed. One explanation for the hiatus can be found in the resources the FWS devoted to endangered species. The agency assigned only two biologists and one secretary to the management and supervision of the entire effort in 1971. The number of professionals tripled the following year, but this level was still intolerably inadequate, or at least many people believed it to be so. A further indication of the program's low priority was the absence of a separate budget line for the program. Until mid-1972, funding for the protection of endangered species was scattered throughout many different units in Interior and the BSFW, and much of this funding was devoted to research and captive breeding. Regardless of its source, even departmental officials agreed that the amount spent on endangered species was far less than could be justified.

A possible further explanation for the paucity of listings focuses on the perception of the BSFW of the number of native species faced with possible extinction. After issuing a list of such species in 1970, the bureau apparently believed it had already identified most of the native species that were endangered. The bureau had certainly listed the species for which there was general agreement, but its appraisal of its own efforts is especially instructive. The BSFW emphasized that its endangered species activities were in a transition between two stages in the early 1970s. The first stage had been devoted to identifying which species were endangered and why. The BSFW explained it this way: "Although many of these questions still exist, many have been answered." [12] These answers led Interior to inform Congress that the number of species already listed as endangered approximated the total number of native fish and wildlife actually in danger of extinction. [13] Accordingly, since most species in need were already listed, the BSFW felt comfortable in moving to the second stage, the development of strategies to remedy threats to endangered species. The bureau did indicate that some additional species would be listed, but many of these would be mollusks and crustaceans, none of which had previously been listed.

Whatever the reasons for the listing hiatus, it did not go unnoticed in Congress. In one report issued in mid-1972, the House Appropriations Committee urged Interior to "become more aggressive in carrying out the intent of the Endangered Species Conservation Act . . . by placing animals on the endangered species list . . . when there is a reasonable amount of evidence to indicate they may be faced with extinction." [14] This encouragement to act was not notably successful, but further statutory action succeeded in getting Interior's attention.

Landmark Legislation, Landmark Change

President Nixon requested in early 1972 that Congress revise the existing endangered species law. Congress made an effort to be responsive to this request, but did not succeed until December of the following year, when the Endangered Species Act of 1973 was passed. The new law repealed the Endangered Species Conservation Act of 1969 and significantly increased the procedural and substantive requirements associated with listing. Whereas the Conservation Act contained only one paragraph on the listing process, the new law included several pages of detailed guidance and several important changes. Many of these changes served as the basis of subsequent amendments, so it is necessary to provide some detail about the implications of the 1973 law.

First, the law increased considerably the range of species that could be listed. Rejecting Interior's preference for vertebrate species, the new law's definition of fish and wildlife was expanded to include any member of the animal kingdom, except those insects determined to be pests "whose protection . . . would present an overwhelming and overriding risk to man." [15] The department was also given authority, which it had not sought and did not want, to list any member of the plant kingdom.

This enhanced jurisdiction was not necessarily a welcome one. The public and politicians alike might be sensitive and responsive to the desirability of protecting well-known animal species; the same could not be said about the likely support for the listing of insect species, like the valley elderberry longhorn beetle, or plant species, like the rhizome fleabane or Canby's dropwort. These are species that most people have never heard of and will never see. Lynn Greenwalt once admitted as much: he agreed that it was "extremely difficult" for him, even among his professional compatriots, "to inspire appropriately serious consideration of the pink-bellied stickleback. . . . This is a disadvantage that we labor under in all quarters." [16]

At the same time, the OES was small, understaffed, and with no recognizable scientific expertise with the millions of species and subspecies of plants and insects. To be responsive to congressional concerns for these species, the OES would have to hire botanists and entomologists. This was not necessarily an appealing prospect in the face of staff shortages in other areas and a preference for vertebrate species. In recognition of the Service's lack of familiarity with plants, Congress directed the Smithsonian Institution to compile within one year a list of native plants that were or might become endangered or threatened and to recommend appropriate action.

Second, the 1973 law retained the existing requirement that listings be based on "the best scientific and commercial data available." This provision was not meant to limit administrative discretion because, as the House report on the law noted, the act was written in such a way as to allow the listing of a species "for any legitimate reason."[17] In addition, however, the law added several procedural steps designed to ensure that affected states and any interested members of the public be given an opportunity to express their opinions about proposed listings. When comments on a proposal were received, Interior would be required to summarize the comments when the final rule was published in the *Federal Register*. The law provided still other opportunities for the public to influence the listing process. Any person who believed that the listing of a species would adversely affect him could request a public hearing. From Interior's perspective, public hearings would be a nuisance and a "costly impediment to the listing process."[18] Consequently, the department opposed the public hearing proviso, albeit unsuccessfully.

Another change in the law allowed any interested person to submit a petition requesting that a species be listed or delisted. If the petition contained "substantial evidence," the Endangered Species Act required that a review of the species be conducted to determine whether listing or delisting would be appropriate. Interior prophetically predicted that this petition procedure would be difficult to implement given the limited staff to examine each petition.

The law did provide one way that the secretary could avoid, at least temporarily, the opportunities for public involvement. When there was a determination that an emergency existed and that it posed a "significant risk" to the well-being of an animal species (and later to plant species), the species could be listed immediately (but temporarily) without regard to public comment for 120 days unless that period was subsequently extended.[19]

Third, the law created two categories of endangerment—endangered and threatened—and removed the requirement that a species had to be faced with possible worldwide extinction before it could be listed. Henceforth, a species could be classified as endangered if it was "in danger of extinction throughout all or a significant portion of its range," or, as threatened, if a species was "likely to become an endangered species within the foreseeable future throughout all or a significant portion of its range."[20]

Making a distinction between endangered and threatened species also had consequences for the kinds of protection available to species in each category. For endangered fish and wildlife, but *not* endangered plants, the protective measures were both prohibitive and comprehensive. Any "taking" of such species would be prohibited regardless of their location on state, federal, or private land, whether it involved hunting, harassing, shooting, wounding, killing, trapping, capturing, or whatever. As the Senate Commerce Committee's report explained it, the prohibition on taking was meant "in the broadest possible manner to include every conceivable way in which a person can 'take' or attempt to 'take' any fish or wildlife."[21] Comparable provisions precluded foreign or interstate commerce in an endangered species, except with prior approval "for scientific purposes or to enhance the propagation or survival of the affected species."[22]

The prohibitions on taking were intended to protect endangered species, but Congress did not extend similar protections to plants. Indeed, the 1973 law treated plants, in terms of taking, in a substantially different manner than other species. Other than prohibiting foreign and interstate commerce in endangered plant species, Congress imposed no restrictions on what could be done to them. This represented a curious distinction between plants and animals. It meant that anyone could take or destroy an endangered plant with impunity, even on federal property, engage in the intrastate sale of such plants, and even move such plants from one state to another except when this movement involved a change of ownership.[23] This discrimination was ironic inasmuch as there was congressional recognition that plants "are even more vulnerable to man's intervention than animals, since [plants] cannot move to more hospitable circumstances when necessary."[24] This logic had convinced at least some members of Congress that there ought to be a prohibition on the taking of endangered plants, and some of the initial legislative proposals contained such prohibitions. In the face of significant opposition from some members of Congress and the Nixon administration, however, these proposals were deleted in the versions that passed in the House and Senate.

For fish and wildlife species classified as threatened, the protective rules could be much different from those for endangered species. A threatened species could be given the same level of protection as endangered fish or wildlife, but the Endangered Species Act authorized the issuance of regulations deemed "necessary and advisable to provide for the conservation" of the threatened species. In other words, there might be limited legal taking of a threatened species, an allowance for inadvertent taking as might occur with fish, or even commerce in the species. The threatened category is useful not only in allowing the listing of species that are not quite on the brink of extinction, but also for previously endangered species whose situation has improved. Having two categories also provides flexibility in that a single species can be listed as both endangered and threatened, depending on the species' status in different parts of its habitat. The dual-listing procedure is occasionally used. The gray wolf and the bald eagle provide some examples of species that are found in both categories.

In addition to introducing important changes, the 1973 act also authorized funding explicitly for the law's implementation. Interior had spent less than $1 million on endangered species in fiscal year 1973. The new law substantially increased this funding, authorizing Interior's expenditure of $22 million for the next three fiscal years (1974–1976). This significant increase indicated that congressional expectations for achievement were quite high. In short, Interior had a mandate for action, a situation that was not at all atypical for much of the environmental legislation approved in the 1970s.

Excessive Expectations or Bureaucratic Inertia?

Congressional expectations for increased listings were high, but these hopes did not ensure either an effective or an immediate response. The 1973 law introduced many new provisions having to do with responsibilities other than listing, and the leadership of the newly reconstituted FWS decided that appropriate rules and regulations were important prerequisites for virtually all other activities, including the listing of species. Consequently, no native species were listed as either endangered or threatened in 1974 and only eight were listed in 1975. Clearly, the congressional goal of preserving endangered species would not be achieved without their being listed.

Agreeing that the pace of activity had been less than expected,

Keith Schreiner, who headed the OES, exercised what can be labeled as bureaucratic accommodation: "Sure, we've been going slow," he admitted in mid-1975, "but I'm trying to avoid the hard confrontation until we've got some firm foundations built in law and precedent. I'd rather avoid confrontation until I'm firmly entrenched and it's harder to blow me out of the water."[25]

While there was surely a need to develop an appropriate foundation before listing additional species, this need was not the only cause of delay. Despite the law's increased authorizations and Congress's increasing generosity in providing appropriations, the OES was vastly understaffed, having only seven professionals working part-time on listings in 1975. Moreover, the Service's initial experience with the Endangered Species Act found that only about half of the species examined actually qualified for listing. Once a species did become a candidate for listing it was still necessary to get administrative approval for the proposed listing from the OES chief, the FWS director, and others before a proposed rule could be published. Each of these opportunities for review and approval represented potential clearance points. As the number of such points increases, the prospects for successful implementation decrease.[26] Until Interior streamlined its approval process in the mid-1970s, at least twelve offices had to review a proposed listing before it was published in the *Federal Register*. Considering its resources and the time required for processing and approval, the OES estimated that it would be able to list no more than fifty species, both domestic and foreign, per year.[27]

This pace might have been acceptable had the OES been able to determine the number of species eligible for listing and to control the listing process. Neither was the case, and this meant that the FWS faced a situation in which it had only limited control over its workload. This is a problem that is highly undesirable from any agency's perspective, and the Service's experience with the number of vulnerable species illustrates the point.

In compliance with the Endangered Species Act, the Smithsonian Institution reported to Congress in January 1975 that nearly 3,200 vascular plant species in the United States were either endangered, threatened, or already or possibly extinct.[28] Although the report had been prepared in a short time, it was viewed as an excellent summary of the status of American plants. The authors of the Smithsonian report expected that when their list reached Congress, the secretary of the interior would immediately thereafter propose the species for listing. Delay, the authors exclaimed, would encourage collection of many species before they became legally protected.[29]

However imperative listing the plants may have been, the OES faced a problem of enormous proportions with the Smithsonian report. The OES did not have the legal authority to issue emergency listings for the plants; such authority was available only for fish and wildlife. The OES also had only two botanists, both of whom had been hired in the spring of 1975, to review and assess the entire list of plants that the Smithsonian claimed to need immediate protection. These dual handicaps limited what the FWS could do. Thus its formal response did not come until July 1975, when it accepted the report as a petition, stating that the OES would review the species to determine which ones should be listed.[30]

A few months later the FWS indicated to a House committee that final action would occur by the end of 1975, but not before the OES had an opportunity to confirm the Smithsonian data. The office did not want to be hasty in its actions, explained its director, Keith Schreiner; he proposed no species-by-species check of what the Smithsonian had done, but he did feel "that the [two] Interior botanists must check the Smithsonian Institution files to verify the existence of the information which qualifies [the plants] legally for the list. . . ."[31] If these botanists judged the data to be adequate, Schreiner said that he was prepared to proceed "without the administrative record normally required." A spokesman for the Smithsonian, one of the report's authors, assured Schreiner that the data were credible and sufficient to justify the listings.

Despite these assurances, quick action was not a result. On 16 June 1976, nearly a year after the FWS announced its intention to review the plant species and seventeen months after the release of the report, the FWS formally proposed that 1,783 plants be listed.[32] In accordance with the law's mandates, the FWS requested comments from anyone who might have an interest in the proposed listings. More than four hundred comments were received, and most favored listing the plants.

The Smithsonian Institution was not the only source of recommendations for species to be listed. By allowing anyone to petition to list a species, the Congress created a massive workload. Within two years of the enactment of the Endangered Species Act, the FWS had received not only the report from the Smithsonian Institution, but at least sixty other petitions as well. The number of petitions was small, but the number of species associated with the petitions—more than 21,000—was not. Most of the petitions involved one or a few species, but some petitioners requested that the OES act on scores or even thousands of species, ranging from the National Speological Society's three requests that 134 crustaceans be listed to a petition from the Fund for Animals recommending action on more than 20,000 foreign species.

In its efforts to be responsive to those who had submitted petitions, the FWS attempted to process them as quickly as possible. Processing the petitions was not, however, the same as acting on them. Many did not include any evidence to justify the requested action. Among those that did include some evidence, it usually supported the petition, but the FWS would have to determine if other, nonconfirming evidence was also available. For every petition that did include substantial evidence, the Endangered Species Act required the FWS to conduct a review of the proposed species to ascertain whether the requested action was appropriate. This could be a time-consuming process, especially when the petition requested action on species that the FWS had not previously considered or on species for which limited information existed. The Endangered Species Act did not impose any deadline on the Service's response to petitions, but this did not prevent some groups from threatening legal action to pressure the FWS to act favorably on their petitions.[33] In several instances these threats were apparently successful because the FWS acted soon after the suits were filed.

Petitions were one way of expanding the Service's perspective about the number of species that might be endangered or threatened, but important changes seem to have occurred within the OES as well. In the early 1970s, the Service believed that only a few hundred or so native species might require listing. With the expanded definition of species in the 1973 law, however, the OES soon realized that many more species were viable candidates for listing. The conservative estimates quickly gave way, and the director of the endangered species program estimated in mid-1976 that perhaps 200,000 species of plants and animals might be listed as either endangered or threatened. As Keith Schreiner emphasized, "When one counts in subspecies, not to mention individual populations, the total could increase to three to five times that number."[34] This was a remarkable prediction for someone who had indicated only a year earlier that his staff was capable of listing only about fifty species per year. The OES had hired an additional biologist to work on listings since Schreiner had made his estimate; even with the additional person, however, if Schreiner's estimate of the number of endangered species was correct, the OES would still need at least 3,500 years and perhaps as many as 17,500 years to complete the listing process. Few government agencies have their workload planned that far in advance.

Although the numbers were overwhelming, the OES vowed not to be intimidated. It would do its best to ensure that species in need would be given attention. The director of the FWS told one congressional committee in the mid-1970s that the Service intended to list approximately 3,000 animals and 4,000 plants, or about 1,400 species a year, before 1982.[35] The FWS never achieved this goal.

The pace of listings did increase in 1976 and 1977, but neither the pace nor the number of listings attracted the approval that the FWS might have expected. Instead, the agency's newfound assertiveness threatened whatever public and legislative support it may have had because of the nature of several of the species that were listed as well as the consequences of these listings. The Endangered Species Act contained a short, innocuous paragraph that required all federal agencies to do whatever was necessary to ensure that any actions they funded, authorized, or implemented did not either jeopardize the continued existence of any endangered or threatened species or result in the destruction or modification of habitats that were designated as critical to the survival of a species. To avoid either possibility, federal agencies were required to consult with the FWS if their projects might affect a listed species.[36] In conjunction with the prohibition on taking, the consultation requirement had the potential to have enormous consequences. (These consequences are discussed in subsequent chapters.) Could the requirement scuttle projects of the powerful Corps of Engineers or perhaps cancel the construction of federally funded highways or other federal projects? Absolutely, declared Schreiner, the law "could stop literally trillions of them."[37]

Schreiner's hyperbole notwithstanding, the fact that the Endangered Species Act might be used to halt the projects of other agencies violated an essential tenet of successful policy implementation. Agencies that want to succeed with their missions should avoid intrusions on what other agencies consider to be their organizational heartland, that area of responsibility agencies claim to be essential to their existence and well-being. Infringement on another agency's heartland is likely to be especially contentious when the agency is a popular one that distributes government benefits. If the FWS did not already know this, it would soon learn as a result of its decision to classify the snail darter as endangered.

The FWS probably would not have listed the darter, a small fish found in eastern Tennessee, except for a petition that the Service received in early 1975 requesting the use of expedited emergency procedures to list the fish as endangered. The petitioners claimed that the Tennessee Valley Authority's (TVA) construction of Tellico Dam, which had started in 1967, six years before the snail darter was discovered, would not only inundate the darter's entire habitat, but probably destroy the species as well. The FWS responded with uncharacteristic speed, perhaps because this was the case the agency selected to establish the precedent that Schreiner deemed necessary for effective implementation of the Endangered Species Act. Within a few weeks of the petition's receipt, the FWS had reviewed it, concluded that it contained "substantial evidence of the existence of a discrete, new species and of danger to its continued existence," and decided to list the darter. In addition, the FWS had also asked the state's governor for his consent to shorten the normal ninety-day comment period so that an emergency listing could be published and requested that the TVA "undertake a thorough review of the effects of the Tellico project, including the present timber cutting on the Little Tennessee [River], upon this species of darter."[38] The TVA was likewise reminded of its legal obligation to take all measures necessary to conserve the species and of the Endangered Species Act's prohibition of any federally authorized activity that would jeopardize an endangered species's continued existence.

The response from the TVA was hardly a cordial one; it had been battling opponents of the dam for many years. Construction had been halted for more than eighteen months in the early 1970s after opponents successfully filed a suit claiming that the environmental impact statement for the dam had been inadequate. Only after completing a revised statement, which two federal courts had found acceptable, was the TVA allowed to recommence construction.

In the authority's view, many of the dam's opponents had little sympathy for the darter and were using the darter and the Endangered Species Act as a subterfuge to prevent further construction. Moreover, the dam was more than half finished and over $50 million had already been spent on construction when the petition was filed. The TVA also pointed out, quite correctly, that Congress and several presidents had approved the expenditure of every dollar for a project deemed to be in the public interest with the full knowledge of the dam's likely environmental consequences. To stop construction because of the darter would be intolerable, the TVA claimed, inasmuch as the agency did not believe the darter to be a new species found only in the Little Tennessee River or that either the construction or the timber cutting would be detrimental to the "so-called snail darter" or to its habitat.[39] A dam would be far more beneficial to the residents of eastern Tennessee than would preservation of a biologically insignificant fish. Upon completion, the dam and the resulting reservoir would provide flood control, hydroelectric power, recreational opportunities, and improve the area's economic vitality, or so the TVA asserted. Under these circumstances, the TVA said that Interior had no basis for issuing regulations to protect the darter or, more important, for suggesting that construction might have to be stopped to preserve the darter.

These objections did not deter the FWS from listing the species as endangered in October 1975, when the dam was about 70 percent complete.[40] The FWS claimed that most of the TVA's objections were without merit—the FWS's evidence indicated that the darter is "genetically distinct and reproductively isolated from other fishes." The decision to list the snail darter was not the last salvo in the administrative battle. The FWS soon designated part of the Little Tennessee River as the darter's critical habitat, asserting that a sixteen-mile section of the river was vital for the species' continued existence. The TVA then petitioned the FWS to void the habitat designation; the TVA reported that the species had been found elsewhere and efforts to transplant it to another river had succeeded,[41] thereby negating the contention that only the Little Tennessee River provided a suitable environment for the darter.

The disagreement between the two agencies might have remained in the administrative arena for several more years except for the persistence of the original petitioners, one of whom was a law professor at the University of Tennessee. With his assistance, one of his copetitioners filed suit in federal district court in early 1976 asking that the dam's construction be halted because it violated the Endangered Species Act. The court dismissed the suit and denied the plaintiffs' request for an injunction.[42] The court agreed that the dam was likely to jeopardize the species' existence. The court did not believe, however, that it was appropriate to apply the law retroactively or to halt a project so near completion, which was scheduled for January 1977.

The court further noted that Congress continued to provide construction funds after the existence of the fish had become known. Refusing to accept the district court's verdict, the plaintiffs turned to the U.S. Court of Appeals, which did grant an injunction and which subsequently reversed the lower court's decision. However the district court might feel about the dam's merits or near-completion, the appeals court said that, since the dam would jeopardize the species' existence, the mandates of the Endangered Species Act had to be followed.[43]

The TVA's reaction was predictable. After having spent over $100 million and ten years' effort, the agency had no desire to abandon the dam, its principles, or Congress's desire to finish the dam.[44] The TVA demanded that it be allowed to appeal the decision to the U.S. Supreme Court. The demand raised a ticklish issue for the Department of Justice, which would be expected to present the government's case to the Court. The Justice Department would have at least two choices to make. First, would it consent to an appeal? Denying the request to appeal would be a victory for the plaintiff but, more important, for the FWS, which had tenaciously pursued the TVA to ensure the enforcement of the Endangered Species Act. Second, which side of the dispute would the Justice Department take? This was an especially contentious issue. The department's choice would reflect not only the federal government's preferences—development versus preservation—but also serve as a test of the executive branch's commitment to the act.

The Justice Department's decisions sent mixed and muddled signals. The department decided that the TVA could appeal, and that the department would work on behalf of the TVA. Despite these choices, Justice also added a note of ambiguity. In its brief to the Supreme Court, the department outlined why the TVA should prevail. In an appendix to the brief, however, the department also attached the diametrically opposed views of Interior. Perhaps the clearest signal of the administration's leanings came from Griffin Bell, the attorney general. To emphasize the importance attached to the case, Bell personally argued the appeal before the Court. When he appeared before the Court, Bell displayed a dead snail darter to emphasize its relative insignificance compared to that of man. Bell's theatrics aside, the Court issued a landmark decision in favor of the snail darter. In *TVA v. Hill*,[45] the Court ruled that the TVA could not finish the dam as long as completion would lead to the darter's demise. According to the Court, the Endangered Species Act's consultation provisions were absolute, and a balancing of competing objectives was not legally possible. No federal action, however meritorious, could proceed if the action would jeopardize a species's continued existence. Put in other words, the Court interpreted the act to mean that endangered species should be protected regardless of cost or the consequences for development.

Other FWS actions also provoked criticism. In November 1977, the Service proposed to list as endangered two species of fish found in Alabama. The state's congressional delegation and its governor soon challenged the proposals, claiming that they were without scientific merit. As one Alabama congressperson later explained, he found it "unfair for the Service to base a proposal on inadequate, substandard biological data and expect to ram it down the community's throat, or force them to find adequate data with which to defend it. It is even more unfair for the Service to disregard biological data which contradicts the assumptions and substance of its proposals. . . ."[46]

This reaction was not the kind that the FWS had anticipated. On the one side, the Service believed that thousands of species should be listed, even in the face of incomplete scientific data. Indeed, the 1973 act allowed the FWS to justify a listing on the basis of "the best scientific data available." For many species, waiting until more complete information was available might mean that they would receive no protection. If forced to provide comprehensive data, which could take years to collect, many species could be lost to extinction because of administrative delays or inadequate budgets. On the other side, because of the FWS's proposals and the Tellico decision, many people realized that the Endangered Species Act could have far-reaching economic impacts. These impacts might be acceptable to save whales and bald eagles, but would the same adverse consequences be acceptable in order to protect species with limited support or popular appeal? One critic of the listing process cast the debate in these terms: "We are not talking about things that grab your heart and soul, the motherhood gut issues of the sandhill crane or the bald eagle. . . . That isn't the issue. It is those totally unknowns that all of a sudden rear their ugly head to stop human endeavor."[47]

Retreat and Reconsideration

When the hearings began in early 1978 to consider reauthorization of the Endangered Species Act, Congress had several concerns. The

FWS had asserted that thousands of species should be listed; by the end of 1977, however, only 202 native species had been classified as either endangered or threatened. The Smithsonian Institution had recommended the listing of nearly three thousand plants, but the FWS revealed its priorities by listing only four of them. In addition, the Service had acted on few petitions. In its defense, it can be argued that the FWS was underfunded, understaffed, and under attack. Most of the well-known and controversial species had already been listed; the most recent proposals had focused on what might be considered as obscure and irrelevant species. Coupled with the Service's combative attitude and the Supreme Court's Tellico decision, Congress was unlikely to provide additional funding without imposing more stringent controls on the FWS. In doing so, however, Congress faced two potentially competing objectives—protecting vulnerable species from the consequences of development while allowing development to occur in spite of its potential threat to these species.

Given a choice most members of Congress would clearly favor the protection of such species as the bald eagle over a highway or a dam. Support was much less likely if an obscure but endangered beetle might prevent the completion of a federal project. Not surprisingly, therefore, one of the major topics of congressional debate in 1978 focused on which species should be eligible for listing. Advocates of protection, primarily environmental organizations, favored broad discretion and chastised Interior for its inactivity. Surely the existing law allowed the department to list more species, or so the environmental groups argued.

Prodevelopment groups and several members of Congress from districts affected by the FWS took a different approach. For these opponents, listing should be severely restricted. Excluded from consideration, according to some opponents, would be species like the snail darter ("You cannot eat it. It is not much to look at. It is a slimy color."); the woundfin ("a useless rascal"); the Colorado squawfish ("a trash fish"); and other species "whose only benefit to man lies in their existence." A common theme in these objections is that an endangered species deserves protection only when it has redeeming value or demonstrated significance to humans. Senator Jake Garn (R.–Utah) complained for example, that the FWS was "looking for some species that maybe has one more rib than some other and exists in one place and nobody really cares whether it lives or not."[48] Garn opposed the listing of such species, but he did express a willingness to support the listing of "some really fine endangered species," presumably those that would never restrict economic development in Utah.

For Senator William Scott (R.–Virginia) acceptable endangered species included only those determined to be of "substantial benefit to [hu]mankind." Scott asserted that people were intended to dominate other species. It made sense, Scott concluded, to segregate those species "which are beneficial to [hu]mankind from those which do not serve any known beneficial purpose."[49] In his view, only the former species should be listed.

Scott's recommendation received little support, but this did not deter other equally extreme proposals. One proposed amendment would have limited protection to species with "economic or aesthetic value to man." Groups such as the American Mining Congress and the National Forest Products Association recommended that all proposed listings be treated as major federal actions, thus requiring the preparation of economic impact statements. These statements, according to the Mining Congress, would discuss the potential effect of listing a species on competition, the productivity of wage earners, markets, consumers, businesses, and federal, state, and local governments. For an agency that was clearly overburdened and well behind its own estimates of satisfactory progress, economic impact statements were among the last things the FWS would welcome. If approved, such a requirement would have made every listing more of a major, time-consuming activity than it already was.

From the perspective of the FWS, two commodities would be most welcome, namely time and additional resources. Substantive and procedural changes could easily disrupt whatever progress the Service had already made, and its experiences with the 1973 law suggested that the FWS would not be able to cope quickly with new legislative requirements. Anything more than modest change could cause significant delay in the listing process. Many members of Congress shared this concern, but they were equally apprehensive about the growing opposition to and potential consequences of the existing law. The Endangered Species Act of 1973 had passed without much opposition, and there is some reason to believe that many members of Congress viewed it as largely symbolic. As noted earlier, the FWS had at least suggested in congressional hearings in 1973 that most species believed to be endangered had already been listed. In fact, however, the Tellico decision and the Service's recent judgment that thousands of species should be listed convinced many lawmakers that the act was far more than a symbolic commitment or, at best, no more than an effort to protect the most deserving species. Senators and representatives alike were under intense pressure from interest groups protesting that the FWS had abused its discretion and that the law was hopelessly flawed because it ignored human and economic values.

Other members of Congress, though supportive of the law, were frustrated due to delays in the listing process. They wanted to see quicker action and more effective implementation. The existing law imposed no time limits on the FWS. Once a species was proposed for listing, the issuance of a final rule might not follow until many months or even several years had passed. Such long intervals caused considerable uncertainty among developers and government agencies that wanted to know whether their projects might affect a listed species.

Whatever the competing claims, Congress opted for change and, prospectively, for substantial modification in the decision rules affecting the listing process. The Endangered Species Act Amendments of 1978 altered considerably the procedural requirements associated with the listing of species.[50] In addition to imposing new requirements for public notice and providing new opportunities for public hearings, the 1978 amendments required that a review of all listed species be conducted at least every five years. The purpose of the review would be to determine whether a species should be delisted or have its status changed from threatened to endangered (or vice versa). This change, although apparently innocuous and surely well intended, would likely increase the workload of the FWS. Not only would the FWS have to review the status of candidate species believed to be endangered or threatened, but the FWS would also have an obligation to ascertain the current status of every species ever listed, including those listed in the 1960s on the basis of educated speculation.

Other changes in the law raised far greater concern. There was widespread recognition that a primary cause of extinction is alteration or destruction of habitats. Despite this recognition, the provisions of the 1973 Endangered Species Act did not adequately address the problem. That act had provided authorization for the federal government to purchase habitats, but the expectation was never that all habitats would be acquired. In some instances, such as when another federal agency already owned a habitat, it would make little sense to spend the limited FWS resources to obtain that habitat.

To address the habitat issue, the 1978 amendments required that when a species was proposed for listing, it would also be necessary, "to the maximum extent prudent," to specify its critical habitat.[51] This habitat was defined as the area the species occupied "on which are found those physical or biological features (I) essential to the conservation of the species and (II) which may require special management considerations or protection."[52] In short, the expectation was that listing a species and designating its critical habitat would be a single, inseparable process. This arrangement, however desirable, would add to the FWS's workload. Prior to 1978, designation of critical habitats had been discretionary. Previous designations of critical habitat had occurred after the affected species had been listed, and in no case had determination of a critical habitat delayed the listing of a species.

The 1978 amendments ensured that this would no longer be the case. While complaining about delays in the listing process, Congress guaranteed that further delay would occur by linking listing with the designation of critical habitats. Critics of the program could readily point to the Service's previous experience with listing species. In fact, the FWS was taking increasingly longer periods to list species once they had been proposed. For species proposed in 1973, less than five months elapsed between the date of the proposal and final action. For species proposed in 1976, however, the average time from proposal to final rule exceeded thirty months. Undoubtedly, the 1973 law's changed decision rules contributed to some of this delay as did the controversy associated with many of the proposals.

Congress also ensured additional delay by requiring an assessment of the economic and "other relevant impacts" of designating any particular area as critical habitat before the habitat was proposed. Based on this assessment, economic criteria could be used to exclude part of a habitat from what was officially designated as critical habitat; this would occur if it was determined that the benefits of such exclusion outweighed the benefits of specifying the area as part of the critical habitat.[53] In other words, economic and not just biological variables would be introduced into decisions to list species. Although these considerations would not prevent a species from being listed, the need to consider economic and other values would add further steps and, therefore, delay the listing process. Moreover, when a critical habitat was proposed, a public meeting would be required. In many ways, as a result, the FWS's flexibility and discretion were reduced considerably. What had at one time been a carte blanche to list species had become a highly constrained and cumbersome process.

In addition to changing the procedures to add species to the list of endangered species, Congress also responded to the controversy surrounding the snail darter. The 1973 Endangered Species Act, which was applicable at the time of the Tellico decision, required federal agencies to ensure that their activities did "not jeopardize the continued existence" of any endangered or threatened species or adversely affect their critical habitats. Attempting to provide the balance that the Supreme Court had indicated was absent from the 1973 law, Congress introduced a process whereby an agency could request an exemption to a jeopardy finding. To summarize briefly a rather detailed and elaborate process, the 1978 amendments authorized an affected federal agency, a license applicant, or an affected state's governor to request an exemption from the no-harm requirement. A review board would first consider the appeal. If the board determined that an "irresolvable conflict" existed, the board was obligated to submit a report to another group. This group would include the secretaries of the army and the Departments of Interior and Agriculture, the chairman of the Council of Economic Advisors, the administrators of the EPA and the National Oceanic and Atmospheric Administration, and one person representing the governor of the affected state. This group could allow a project to proceed, but only if a series of stringent criteria were met. Exceptions could be granted if the committee concluded that the project was of regional or national significance and that there were no reasonable or prudent alternatives to its completion.[54]

There is no doubt that the exemption procedure was intended to mitigate any future situations like the Tellico controversy. Indeed, the law specifically required the committee to consider exemptions for the Tellico Dam as well as for another project, the Grayrocks Dam and Reservoir Project in Wyoming, that had also produced controversy because of claims that it would affect the habitats of whooping cranes. The exemption provision had been added at the instigation of Senator Howard H. Baker, Jr. (R.–Tennessee), a strong advocate of the Tellico Dam and the Senate minority leader.

When the appeals procedure was added to the 1978 law, many environmentalists feared that it would lead to a large number of exemptions and intentional decisions to extirpate many species. Despite this concern, the full process has been used only for the two projects specifically mentioned in the law. In the Grayrocks case, the committee unanimously accepted a compromise agreement that allowed completion while protecting the cranes' habitats. In contrast, the committee unanimously concluded that reasonable alternatives to the Tellico Dam existed and that the benefits of preserving the snail darter outweighed those associated with the dam.[55] Senator Baker was reported to have been furious with this decision. His intended solution had failed in a rather embarrassing way. Baker had lost the battle, but he remained in the war. He introduced legislation to abolish the committee and exempt the Tellico Dam from all provisions of the Endangered Species Act. He was not able to abolish the committee, but his persistence and deft legislative manuevering allowed completion of the dam. With his strong support, Congress passed the Energy and Water Development Appropriation Act of 1980, which directed the TVA to finish the dam, "notwithstanding . . . any other law. . . ."[56]

Cecil D. Andrus, the secretary of the interior, urged President Carter to veto the bill, thus preventing further construction. Carter disregarded this advice and signed the bill. Construction was restarted and the dam was soon finished. Many environmental groups were dismayed that Carter had accepted the dam and the snail darter's possible extinction. Ironically, additional biological surveys subsequently identified several more darter populations, and the darter was reclassified to threatened in July 1984.

The darter controversy was also indicative of more important change. When Congress passed the Endangered Species Act in 1973, it gave Interior considerable discretion. Used wisely and effectively, this discretion could have allowed the FWS to achieve many of its objectives in the listing process. If a few powerful interests viewed the use of this discretion as arbitrary, however, then congressional intervention would surely follow. Indeed, it did. The Endangered Species Act Amendments of 1978 severely constricted the administrative latitude that the FWS had once enjoyed. No longer could species be listed with only minimum publicity and attention. Similarly, if some listed species could cause problems for other federal agencies, then the congressional defenders of these agencies would intercede to protect their interests.

The 1978 amendments constrained what the FWS could do, but the impression should not be left that the amendments permanently crippled the listing process or other efforts to protect endangered species. Questions had been raised about how the FWS administered the program, but this did not diminish congressional support for the idea of protecting vulnerable species. Congress was overwhelmingly in favor of protecting them. Only three of ninety-seven senators voted against the 1978 amendments. In the House, a similar pattern was evident; of the nearly four hundred votes cast on the amendments, only a dozen representatives did not support them. Few federal agencies can point to such support. In short, there was reason to expect continued support for the listing process, but only so long as the FWS administered it in an effective and noncontroversial manner.

he changes associated with the 1978 amendments meant that
the OES would once again have to revise its procedures, prepare new regulations addressing the listing process and the
designation of critical habitats and, for the first time, develop
a competence in economic assessment. Congress had overwhelmingly
approved the changes, but this did not ensure that the FWS would be
able to translate its legislative mandate into effective action. In fact,
what happened soon suggested that expectations for such action were
entirely unrealistic.

The proof of this occurred quickly. The requirements of the 1978
law had an immediate effect on the listing process. In March 1979,
a few months after the law's passage, the FWS announced that it was
withdrawing proposals to designate more than seventy critical habitats until new regulations were issued and the required economic
reviews were completed. Perhaps more important, in order to apply
the law's new procedural and economic impact requirements, the FWS
said it would voluntarily delay the already-proposed listing of over
150 animals and more than 1,800 plants, most of which the Smithsonian Institution had recommended for listing in early 1975.[1] These
voluntary withdrawals concerned many people, especially because of
Congress's establishment of deadlines for listing species.

To complicate matters, the 1978 amendments mandated that a final
rule listing the species had to be published within two years for species
proposed for listing *after* the date of the law's approval. If the deadline
was not met, the proposed listing had to be withdrawn. For species

proposed for listing at least two years *prior* to the law's approval, all rules had to be finalized within one year (that is, by 10 November 1979) or be withdrawn.

Could the FWS meet the new procedural requirements *and* issue final rules for over 1,900 species within eight months? The FWS initially believed so and emphasized that it was "committed to providing full protection as quickly as possible for endangered and threatened species. . . ."[2] In spite of this assurance, the apprehension that existed elsewhere was well founded. Between December 1978 and November 1979, the FWS listed only thirty-four species, all plants. Moreover, most of these plants were listed within days of the 10 November deadline. When the deadline came, the FWS formally withdrew proposals to list 1,876 species: 63 vertebrates, 87 foreign plants, and 1,726 native plants. The large number of plant species left unprotected was especially disconcerting. Among the native plants, the Smithsonian Institution had recommended in January 1975 that all should be listed and protected as soon as possible. Nearly five years after receiving the recommendations, however, the FWS had listed only fifty-six native and one foreign plant species—or about 3 percent of the Smithsonian's list. Among the plants listed, several did not face a high threat while others that did face a high threat were listed as threatened rather than endangered.

In justifying the withdrawals, the FWS cited not only the requirements of the 1978 amendments, but other procedural burdens as well. Concerned about the economic effects of government regulations, President Carter had issued an executive order requiring all federal agencies to assess the economic and record-keeping impacts of proposed rules on local, state, and federal programs. Similarly, Interior decided that all proposed rule-making packages had to contain impact assessments outlining the environmental consequences of listing plants and animals.[3] In the face of these procedural barriers, the FWS claimed that it would have been a "miraculous feat" to have listed all the proposed species before the November deadline. What the FWS did not point out was that it had proposed most of the species for listing more than three years earlier.

Environmental groups were completely unsympathetic to the Service's defense. Several charged that bureaucratic incompetence provided the best explanation for what had occurred. At least one scientist from the OES agreed, conceding that the inaction constituted administrative "incompetence and ineptitude." His superiors, he said, were "afraid to take the bull by the horns because they worried what would happen politically to the Endangered Species Act itself."[4]

Although the FWS could repropose the withdrawn species, the new law guaranteed that this would not soon occur. The law stipulated that once a "delinquent" proposal was withdrawn, a species could not be reproposed for listing unless there was "sufficient new information . . . to warrant the proposal of a regulation."[5] The FWS interpreted this provision to mean that the "new information" would have to be obtained *after* the withdrawal rather than after the initial proposal. This was an important distinction because it precluded the use of information gained during public hearings or provided as a result of the initial listing proposal. The consequence of this requirement puts its effect into some perspective. When the FWS proposed, in 1976, that approximately 1,800 plant species be listed, the Service presumably had enough information to warrant their listing. Because of the "new information" requirement, however, among the eighteen hundred species, only four were reproposed and listed as endangered or threatened in the two years following passage of the 1978 amendments. Later amendments to the Endangered Species Act deleted the "new information" requirement, but less than 120 native plants were listed in the eight years between 10 November 1979, when the plant proposals were withdrawn, and 31 December 1987.

The release of a highly critical report in mid-1979 prompted additional concern that considerations other than biological ones affected FWS decisionmaking processes. In *Endangered Species—A Controversial Issue Needing Resolution*, the GAO reported that the Service's implementation of the listing process was remarkable for its deficiencies. The GAO concluded that these deficiencies threatened the effectiveness of the entire endangered species program.[6] Specifically, the GAO found problems in three key areas related to the listing of species: the variables included in decisions to list species; the handling of petitions; and the priorities that the OES used to determine which species would be listed first. Due to the seriousness of the GAO's concerns, each of these topics deserves some attention.

Biological Versus Nonbiological Considerations
as Determinants of Listings

Previous chapters stressed the difficulties associated with obtaining definitive scientific data about species in jeopardy. Despite this problem, government agencies have a natural preference to justify their policies on the basis of empirical observations and on evidence believed to be objective, compelling, and (relatively) indisputable. The FWS is not an exception. The Service maintains that its decisions to list species are based on "the best scientific and commercial data available," just as the law requires. To admit otherwise would impair the credibility of the FWS and provoke justifiable complaints about species already listed or proposed for listing. While acknowledging that its decisions are "necessarily predictive" and ultimately dependent on the competence of its scientists, the FWS has frequently emphasized the objectivity of the entire listing process as if it is a rational one, independent of the politicized environment in which the agency operates.

As an assistant secretary of the interior, Robert Herbst, once explained to a congressional committee, the listing of a species is a "pure, professional, objective, biological opinion." A fellow OES spokesman, Earl Baysinger, agreed and noted that such nonbiological considerations as controversy are irrelevant: "[I]t makes little difference whether or not the species is controversial—we intend to list them when they qualify regardless and . . . controversy has little or nothing to do with the time and effort it takes to list a species—the same basic data and the same set of documents are needed in either case."[7]

Many people disagree with these assertions and claim that nonbiological variables frequently explain what the FWS does. To support their case, critics point to statements from FWS employees. As an illustration, when asked in 1975 to provide a list of species for which the Service had received petitions, the FWS noted, as if controversy was a relevant consideration, that it was attempting "to identify those taxa which could be listed without creating major controversies."[8]

Far more important than these contentions is an assessment of what the FWS has actually done when it has listed species. This assessment shows that nonbiological variables, such as controversy and political concerns, do provide an explanation for many listing decisions. To begin, the law specifies five criteria for listing and imposes a nondiscretionary obligation to list a species whenever it faces any of the following problems:

1. The present or threatened destruction, modification, or curtailment of its habitat or range
2. Overutilization for commercial, sporting, scientific, or educational purposes
3. Disease or predation
4. The inadequacy of existing regulatory mechanisms, or
5. Other natural or manmade factors affecting its continued existence[9]

Using these criteria as a starting point, the GAO attempted to determine whether they were, in fact, the criteria actually used in the listing process. In two of the areas it examined, the GAO discovered considerable inconsistency. In the first area it found that the criteria were relevant to the process, but so also was the potential for conflict with other federal agencies. In one case, the GAO pointed out that in mid-1972 the FWS was aware that the construction of a TVA dam on the Duck River in Tennessee might jeopardize the existence of a species of mollusk. Even with this knowledge, however, the FWS did not list the species until mid-1976, after the dam's completion, at which time none of the mollusks could be found.[10]

This finding suggested the desirability of assessing FWS procedures used to determine how other federal water-related projects might affect the status of species under consideration for listing. The GAO found inconsistencies and several instances in which the procedures used to list controversial species differed from the procedures used to list noncontroversial species. In every project discussed in its report, the GAO noticed a discrepancy between what the OES's biological data warranted and the action eventually taken.

The Illinois mud turtle provides one of the more prominent examples of apparent political influence in the listing process. The turtle, a relict subspecies, is found in Iowa, Illinois, and Missouri. The FWS proposed that the turtle be listed as endangered in mid-1978 because its habitat was "subject to intense alteration and collection of individuals [was] a threat to the continued survival of this turtle."[11] The Service also noted that the dumping of poisonous chemicals in ponds probably contributed to the species' precarious position. The judgments about the turtle's status were based on what the FWS considered to be the best available biological data, and the proposed listing received support from the Missouri Department of Conservation and the Illinois Division of Wildlife Resources.

The species' habitat includes part of the facilities of the Monsanto Agricultural Products Company in Muscatine, Iowa, so the company hired a consultant to conduct an independent assessment of the species' status. This assessment concluded that the mud turtles are a separate population of common turtles, that the mud turtles were far more common than the FWS believed, and that the available scientific data did not justify listing the species, just as one might have expected considering the source of funds for the study. As Monsanto claimed, the FWS proposal "was based on insufficient and poorly documented scientific data." When presented with more appropriate data, according to one of Monsanto's officials, Interior "continued to use the most shabby technical information, supposition, speculation, and innuendo." [12] After receiving the consultant's report, the FWS arranged for its review by several turtle specialists. The eight scientists who responded severely criticized the report.

The dispute changed from a scientific to a political one when Monsanto succeeded in getting Iowa's U.S. senators to intervene on its behalf. Both complained to Interior, and Senator Roger W. Jepson (R.–Iowa) urged Secretary of the Interior Cecil Andrus to reassess the proposal to list the turtle in order to avoid "an arbitrary and capricious ruling." [13] Interior accepted a suggestion that it obtain an external review of the scientific data. This review concluded that the species was in need of protection. Although the OES had already begun preparing a final rule on the turtle, Interior withdrew the proposed listing, stating that additional research was necessary before any action would be taken.

Insufficient scientific data also provides explanations for other decisions not to list species that face a high degree of threat. In the 1970s the FWS delayed the listing of several species of spiderlike invertebrates that could be eliminated because of the construction of a dam in California. The delay occurred even though the OES had concluded that the species were endemic to the area, that fieldwork data were adequate, and that the dam would cause the species' habitats to be inundated with water. The Service nonetheless claimed the data necessary to support the listing was insufficient, so no action was taken.[14]

In other situations, however, species not in conflict with development projects were listed even in the absence of detailed status surveys. For one species that was listed, the Pine Barrens treefrog, the FWS director agreed that the field notes of one person "were (and are) the only source of information on the distribution and range of this species available." [15] The GAO implied that the treefrog should not have been listed, but the FWS maintained otherwise; the Service replied that those who wrote the GAO report "did not take the time to familiarize themselves with the biology of the Pine Barrens treefrog nor contact those who are familiar with its status." [16] Having accused the GAO of incompetence, the FWS later recanted, admitted that its original data on the frog were in error, and delisted the species in late 1983.

The GAO asked the FWS to provide an explanation for the apparent inconsistencies in some of its actions (e.g., delaying the listing of the spiderlike invertebrates despite the availability of appropriate scientific data, listing the Pine Barrens treefrog despite having only limited information). The manager of the endangered species program defended these actions in this way: "Not all species can be saved, it's a judgment decision, somebody has to play God and decide which will go. In this case I am doing just that, contrary to my staff's recommendations. I am making biopolitical decisions every day. Right now my main concern is saving the act. . . . I am not going to lose the act because of a couple of spiders. Some species will have to become extinct." [17]

Controversy has also played a role in the administrative processing of listing proposals. The FWS had been delegated the authority to issue final rules listing species as endangered or threatened, but it was expected to submit all potentially controversial regulations to the secretary of the interior for his personal review and approval.[18] When he headed the OES, Keith Schreiner once called attention to this expectation in a congressional hearing about grizzly bears: "I cannot tell you that the final determinations are based only on biological evidence, because you know as well as I do that other considerations—political, economic, and the like—enter into such matters. . . . The final decisions rest with the Secretary, and it is at that level where other considerations can enter into the decisionmaking process." [19]

These comments can be interpreted in terms of the number of clearance points discussed in the previous chapter. Controversial proposals have more clearance points than routine proposals and more opportunities to either delay or impede the listing process. It would be unusual if this were not the case: controversial political issues rise to higher levels of government to receive attention not just from civil servants or career program administrators, but also from elected and appointed policymakers who are far less likely to be attentive to long-term biological considerations than to short-term political repercussions.

Indeed, the potential repercussions may discourage the listing of potentially controversial species. Just such a situation occurred with the Northern spotted owl, a species that inhabits the forests of the American Northwest. Although scientific evidence suggested that the owl's existence is in jeopardy, listing the species would curtail the timber industry in Oregon and Washington, or so industry spokesmen claimed. After the FWS decided not to list the owl, allegedly because of these claims, environmentalists filed suit. They argued that the Service had ignored the scientific data and had violated its legal obligation to protect a species in danger. Ruling on the case in late 1988, a federal judge declared the Service's decision to be "arbitrary, capricious, and contrary to the law." An investigation of the episode found that the scientific data justified listing. Despite this, the agency's management officials had not only overruled their own scientists but had also altered substantially the scientific evidence to understate the threat to the owl.[20] With these findings, the FWS had little choice; the owl was proposed for listing in mid-1989.

Not only political but administrative considerations also seem to affect the listing process. Some evidence suggests that scientists within the FWS have been informally evaluated on the basis of their productivity, that is, how many species they have listed, rather than on the number of controversies with which they have become involved.[21] In other words, the Service's reward structure can skew the listing process in certain directions, to the detriment of species prone to conflict, which often take longer to list than is the case with routine listings.

The concerns of the GAO were not limited to the role of political or administrative considerations. It also examined how the FWS decided to classify species as either endangered or threatened. Having two categories increases the Service's discretion, but the Endangered Species Act provides limited guidance about how to classify species in danger of extinction. Relative threat to existence is supposed to guide these decisions, but the perception of threat can change or vary among biologists responsible for listing different classes of species. As a result, such biologists recommend either endangered or threatened status on the basis of their professional judgments rather than on the strict application of standard guidelines.[22]

A grandfather clause in the 1973 law further compounded the

Table 6.1   Degree of Threat to Species Listed as Endangered Prior to
1974 and to All Species Listed as Threatened, 1974–1986 (in percentages)

Degree of Threat	1978		1983		1986	
	E	T	E	T	E	T
High	38.5	4.0	43.2	22.4	52.9	21.2
Medium	49.0	68.0	45.2	67.4	36.5	68.8
Low	12.5	28.0	11.5	10.2	10.6	10.0
Total	100.0	100.0	99.9	100.0	100.0	100.0
(N)	104	25	104	49	104	80

Source: Computed from data provided by the FWS.
Note: The endangered category includes only those species listed prior to 1974 that
were so categorized in each of the three subsequent time periods.
Key: E = Endangered, T = Threatened

problem in making an appropriate distinction between the two categories. According to the law, all species listed as endangered before
its passage would retain that designation, regardless of the threat
they faced, unless the FWS changed their status from endangered to
threatened or delisted them. For many of these species, classification
as threatened probably would have been more appropriate, except of
course that such designation was not available prior to 1973.

With the availability of the threatened category, the FWS could have
reviewed the grandfathered species. The Service decided, however,
to forego this opportunity in the belief that it was more important to
focus its limited resources on deserving but as yet unlisted species. As
a consequence, many of the species listed prior to 1973 should not be
listed as endangered if one considers the degree of threat they face.
Likewise, many endangered species are not in immediate danger of
extinction (see table 6.1). By way of comparison, table 6.1 also shows
that there are many threatened species that face a high degree of
threat and are in jeopardy of extinction. In short, the FWS's own data
reveal that some endangered species face a low threat to their survival, and some threatened species face a high threat. This pattern is
just the opposite of what Congress intended.

The degree of threat to a species is supposed to determine its listing
as endangered or threatened, but there is further evidence that other
variables affect the choice of categories. When a species is proposed
for listing, there is an implicit expectation that the FWS has data of
sufficient scientific merit to justify its proposal. Furthermore, before

publication in the *Federal Register* as many as a half dozen officials will have reviewed a proposal, and they have to be satisfied with the scientific justification before publication of a proposal to list a species. Only in rare instances, then, would one expect information obtained subsequent to the proposed listing to affect this choice. There might be some instances, however, when the threat to a species diminishes, when additional populations of a species are located, or when public hearings provide new scientific evidence about a species' condition. In any case, one can hypothesize that: a) once a species is proposed as endangered or threatened, the final categorization will be the same; and, b) the number of changes (from proposal to final rule) from endangered to threatened or vice versa will be approximately equal.

At least the first hypothesis is correct. Considering that nearly six hundred native species have been listed, there have been less than twenty-five instances in which the final categorization differed from the one proposed. In contrast, in virtually all changes, a species was proposed as endangered but eventually listed as threatened, the less restrictive category. For some changes the explanation is straightforward, as with five species of plants in the Ash Meadows area of California. Between the proposed and final rule, the FWS purchased the species' habitat from a real estate developer, thus diminishing the threat to that habitat.

In other instances the biological explanation is much less certain. Most of the species involved are plants, reptiles, and snails, species that engender limited public support. In at least eleven of the sixteen cases, a state or federal agency opposed the listing as endangered.

Opposition from influential politicians similarly provides an explanation for some changes, and Utah's desert tortoise offers one of the best examples. For millions of years the tortoise has inhabited parts of what is now southwestern Utah. Studies conducted in the 1930s and 1940s found about two thousand tortoises in the only area in Utah in which the tortoise occurs naturally. By the late 1970s, only a few hundred remained on land the BLM controls. The BLM leases the land for livestock grazing, and sheep and cattle compete with the tortoises for food. Due to their declining numbers and habitat, the FWS proposed listing the tortoises as endangered in 1978.[23] The proposal generated a firestorm of controversy. Utah's two U.S. senators requested a public hearing to oppose the proposal. Opposition also came from local residents, the governor, Utah's Division of Wildlife Resources, one of the BLM area offices, and a state senator who accused the FWS of being more concerned with the well-being of "useless animals" than it is with the residents' livelihood.[24] The desert tortoise was eventually listed as threatened, but only at the last legal minute. Had the FWS waited only four more days, the Endangered Species Act's two-year rule would have required the Service to withdraw the proposal until new information was obtained.

Other examples are available, but the evidence presented reveals susceptibility to external political pressures in decisions to list species. Other evidence is similarly available that nonbiological considerations affect the FWS's handling of petitions to list, delist, or to change the classification of species.

The Right to Petition

The 1973 Endangered Species Act, as noted earlier, allows anyone to request or petition that a species be listed. For petitions that include "substantial evidence," the law requires Interior to conduct a review of the relevant species' status. Examination of how the FWS has responded to petitions provides some indication of the agency's reaction to public participation.

Petitions provide an advantage in that they offer an external source of information for the FWS, albeit an inconsistent one. Some petitions are well documented, others are not. Regardless of the merits of the individual petitions, what the GAO found was consistent with its findings about listing procedures—disarray and arbitrary decisions. The FWS had received more than 150 petitions through mid-1978, but not until that year did it establish a procedure to record a petition's receipt or disposition. The result, as might be expected, was administrative chaos and the loss of many petitions.[25] Of equal concern was that what constituted an acceptable petition was not well defined. Even after criteria were formulated to assess the substantiveness of petitions, the criteria were not consistently applied. The Service's biologists frequently made individual judgments about a petition's merits, often on the basis of their personal evaluation of the data and occasionally on the basis of the petitioner's presumed credibility.

Some petitions led to status reviews in the absence of appropriate scientific or commercial data. Other petitions were denied a review in spite of such evidence. Louisiana, concerned about an increase in the number of alligators, asked the FWS in early 1974 to change the species classification from endangered to threatened in parts of

Table 6.2   FWS Responses to Selected Petitions

Species	Action Requested	Evidence Accompanying Petition	Action Taken
Florida swallowtail butterfly	List as E	None	Listed as T
Smith's blue butterfly	List as E	News articles	Listed as E
Bolson tortoise	List as E	Could not be verified	Listed as E
Stock Island snail	List as T	Literature provided, but not considered a petition by FWS biologist	Listed as T
Asplenium adiantum	List as E or T	Telephone conversation considered as a petition	Status review conducted; not listed
Ferruginous hawk	List as T	Petition recorded, but could not be found	None

Source: House Committee on Merchant Marine and Fisheries, hearings, *Endangered Species*, 96th Cong., 1st sess., 1979, 324–40.
Key: E = Endangered, T = Threatened

the state. After reviewing the petition, a Service biologist concluded that much of the state's data did "not in the slightest give any indication of the present status of the alligator within the state and as such should be disregarded."[26] Just the opposite occurred, the petition was accepted, and the FWS reclassified alligators in three parishes in Louisiana. The data in table 6.2 reveal that such action was not unusual; the FWS initiated changes in status in the face of limited or no evidence. In other cases, no action could be taken because petitions had been lost or misplaced.

California's Department of Fish and Game had a notably different experience. It petitioned the FWS to reclassify the San Joaquin kit fox from endangered to threatened and provided five studies to justify the request. The acting chief of the OES accepted the studies as providing "substantial information" and noted that the FWS regional office in Portland, Oregon, had also recommended a change in status. More significant, the chief agreed that had the threatened classification been available when the fox was listed in 1967, the fox would have been listed as threatened because it was not in immediate jeopardy of extinction.[27] None of these reasons proved persuasive, and the FWS denied California's request.

These and other similar experiences led the GAO to conclude that substantial reform was needed in the handling of petitions. To do otherwise might preclude some species from receiving the protection their status justified. After several attempts the FWS was able to devise an appropriate mechanism to handle petitions, and it is probably not now subject to the criticisms that it once received on this topic.

Spending Scarce Resources

Scarce resources inevitably require government agencies to make choices about how to spend their money in order to satisfy competing claims. This allocation process is especially important when the available resources come from external sources, such as Congress, and when the recipient has only limited control over its workload as does the FWS. In the first instance, an administrative agency can ill afford to neglect the wishes of those who control its purse strings. In the second instance, the biological environment for which the FWS is responsible is rarely static. Some species are suddenly endangered because of natural events, human activities, or for reasons largely unknown. Biologists within the FWS may prefer the application of scientific criteria to the allocation process, but program administrators who ignore social, economic, or political values do so at the expense of their agency's well-being. These considerations make it even more difficult to decide how to allocate resources to the listing process. These difficulties notwithstanding, some kind of priority system is imperative if scarce resources are to be used effectively. Such a system ideally would be practical, parsimonious, and easily understood. From a biological perspective, the system would also ensure that the most vulnerable species will be listed first.

Having identified desirable features of a priority system, the Service's experience quickly reveals the difficulties in devising an effective one. Beginning in the early 1970s, the FWS initiated efforts to devise a system to guide decisions about listing species. At least six separate systems had been considered by 1978, but none had been formally adopted. The apparent explanation was that the FWS "could not agree on the scope, comprehensiveness, criteria (and the emphasis given each), definitions, and other components which should be included."[28]

One of the first systems, developed in the early 1970s, focused on the plight of a species, its importance to ecosystems, and the prospects of success in saving it.[29] Ten separate variables were used to determine an overall score for each species under consideration. These variables included such things as the species' population levels and status, gene contamination, and taxonomic distinction. A species could also gain points if it had some utility, that is, if it had some symbolic value, consumptive potential, or a chance that it might serve a scientific purpose.

A second system developed for the FWS in the mid-1970s had four categories of variables: attributes of the species, its vulnerability, recovery potential, and current population status.[30] The system was far from parsimonious, however, because it required users to make judgments about eighteen distinct variables and then to assign points for each variable for each species. Species with the most points were deemed to be in most need of protection. The points assigned for each variable were, however, somewhat arbitrary. Substantial amounts of information were required to provide an accurate total score. Even when information about a species was relatively abundant, different people interpreted the same data differently, thus producing disparate rankings for the same species. For many species, reliable information was not readily available, so the scheme reflected a cautious approach. Species about which little was known received the highest scores. The assumption here was that a species on the verge of imminent extinction ought to be protected regardless of how little other information was available about the species. Last, the scheme gave no preference to different vertebrate taxa. All were deemed equally important, regardless of their potential appeal or utility to humans.

Still another system was far simpler, and it relied on only three variables: the degree of threat to a species, the availability of information, and whether the plant or animal was a species or a subspecies. The degree of threat facing a species was the key variable. Included in the high threat category would be those species facing a precipitous population decline or an imminent threat that would destroy all or a major part of its habitat. For such species, extinction was "almost certain in the immediate future unless rapid measures" were taken to list the species.[31] Medium threat species faced a continual population decline or a short-range threat to their habitat. Listing could be temporarily deferred without resulting in extinction. In the low threat category would be found rare species undergoing a short-term population decline or a self-correcting fluctuation. The species' habitat would face no known harmful effects within the next five years.

This priority system, which the FWS was developing when the GAO conducted its evaluation, had several appealing features. Most apparent, few decisions had to be made since there were only twelve different priorities. After determining whether the candidate was a species or a subspecies and whether additional fieldwork was necessary, only the degree of threat had to be ascertained. As with the earlier systems, the "new" system also had scientific merit because it did not discriminate among taxonomic groups. Highly threatened insects *and* mammals would be equally likely to be listed, all else being equal. Last, the director of the FWS assured Congress that the OES would allocate its program funds in "strict adherence" to the system.[32]

Regardless of its merits, the system was not without critics. The degree of threat is an important consideration in deciding to list a species, but so also is the potential for recovery, which the system ignored altogether. Moreover, since the listing process is expensive, why list a species with dim prospects for survival when the same funds could be better spent on listing a "recoverable" species or on the recovery of a species previously listed? Similarly, why list one species that will be costly to recover when the money could be spent more effectively on the recovery of several species?

The system's seeming simplicity also created another disadvantage. To be effective, a priority system should resolve disagreements about how to allocate resources. When too many species are included in too few categories, then the system does not indicate which species should be listed first within each category. This would be a particularly acute problem if the number of species in the four high-threat categories exceeded the number of species that could be listed within the time period the species were expected to survive without the benefits of listing.

Exactly such a situation occurred when the FWS first used this priority system. One hundred and forty species were assigned to the highest listing priority. The consequence was that criteria other than those incorporated into the system had to be used to decide which species to list. Using this priority system in mid-1978, the FWS ranked 398 species that it intended to list or anticipated listing before the end of 1979. All priority 1 species would be listed before those in all lower categories, or so the ranking system indicated. In fact, however, this did not occur, as the data in table 6.3 reveal. The data indicate the number of species in each priority category (as of August 1978) and how many of these species were listed in subsequent years.

Table 6.3   Application of the FWS's 1978 Priority System for Listing Species

Degree of Threat	Priority	No. of Proposed and Candidate Species	No. Listed 1978–81	No. Listed 1982–85	No. Listed 1986–89
High	1	140	20	13	15
	2	38	8	4	1
	3	55	0	4	6
	4	8	1	1	2
Medium	5	51	5	4	5
	6	20	0	2	0
	7	49	4	1	7
	8	12	0	0	3
Low	9–12	25	0	0	1
Total		398	38	29	40

Source for priority rankings: House Committee on Merchant Marine and Fisheries, hearings, *Endangered Species—Part 2*, 95th Cong., 2nd sess., 1978, 1009–53.
Note: The table excludes species subsequently delisted.

In examining the data in table 6.3 several things are immediately evident. First, the FWS has not achieved its goal of listing all the species it said were facing a high threat of extinction in 1978. Given the definition of high threat, the ability of the FWS to list only 31 percent of the species with the four highest priorities means that some species have become extinct as a result of their failure to be protected. The Louisiana vole, the pallid beach mouse, and Sherman's pocket gopher represent three such species. All three species faced a high threat in 1978. None were listed, and all are now believed to be extinct.

Second, many species facing a medium threat were listed before all species believed to confront a more serious threat. Of course, one must consider that the situation is a dynamic one and that threats change over time or were incorrectly identified initially. What the table does not show is that most of the high-threat species have never been proposed for listing, at least through 1989. In short, the priority system that the FWS employed (and developed in response to the GAO inquiry) in the late 1970s did not serve as an effective guide to action. Criteria other than the relative degree of threat apparently guided most listing decisions.

Conservatism and Conservation: The Reagan Revolution

Regardless of how poorly the priority system was implemented, by the end of the Carter administration in January 1981, 51 native species were classified as threatened and another 287 as endangered. The 1978 amendments to the Endangered Species Act had frustrated expected progress, but the Carter administration could claim credit for having added over one hundred native species, including more than sixty plants, to the endangered species list.

With Ronald Reagan's inauguration, much would change in the efforts to list species. Reagan's views on the federal government's role are well known. Too much government is an unwelcome intrusion into Americans' life-styles, or so Reagan believed. Not unexpectedly, James Watt, Reagan's choice for secretary of the interior, shared this view about government. The change in administrations had a quick and dramatic effect on the listing process. In the last year of the Carter administration, nineteen species had been proposed for listing. During the next twenty-four months, the FWS produced only six proposed listings. The FWS listed fifty-four species during the last two years of the Carter administration. Over the next two years, only eleven species were listed, and the first of these did not occur until February 1982.

Explanations are readily available for the dramatic decline in the number of listings. Although there is no evidence to indicate that the FWS received explicit directions to slow the pace of listings, other decisions guaranteed that few proposals or final rules would emerge. On the one hand, Interior introduced a change of emphasis. Complaining that too much emphasis had been placed on listing at the expense of appropriate recovery actions for species already listed, the FWS said that it would henceforth devote more of its attention to the development of recovery plans (see chapter 10).

On the other hand, and perhaps more important, soon after taking office, Reagan ordered that most federal regulations not yet in effect be suspended until their economic consequences could be assessed. Final rules to list species do not take effect until thirty days after their publication in the *Federal Register*, so Reagan's decision affected four listings for which final rules had been issued in the last week of the Carter administration. The Reagan administration suspended the effective date of the four listing rules six times. Only when threatened with legal action did Interior relent and allow the listings to take effect.[33]

In February 1981, the president also issued an executive order that required all regulatory agencies to consider, to the extent permitted by law, the potential costs of their proposed regulations and to ensure that the potential benefits to society outweighed these costs.[34] In other words, economic criteria would have to be a consideration in all future decisions to propose a species for listing. Combined with other procedural requirements, considerable delay faced any effort to list additional species. After sufficient scientific information had been obtained to justify a proposed listing, a Determination of Effects (DOE) would have to be prepared in order to comply with the Regulatory Flexibility Act of 1980, the Paperwork Reduction Act, and the newly approved executive order. For each proposed listing a DOE would have to analyze "the significant national or regional impacts, the significant impacts on State or local governments, the significance of new information collection, the major conflicts with other Federal programs, the significant impacts upon industry, and the possible adverse effects upon competition, employment, investment, productivity, innovation, and the ability of domestic enterprise to compete with foreign-based enterprise."[35] After review within the FWS and Interior's Office of the Solicitor, an approval would be necessary from the department's assistant secretary for fish and wildlife and parks, a political appointee. Upon this approval, a proposed rule could be developed, but it first required review and approval from the OMB before publication in the *Federal Register*. The OMB would provide still another review before a final rule could be published. As one might imagine, these new requirements were quite burdensome for an agency that had limited familiarity with economic analysis.

Interior denied that economic variables would determine whether a species would be listed,[36] but the OES chief, John Spinks, was not so sure. He complained in 1981 that the inclusion of economic variables in the listing process went well beyond what the Endangered Species Act allowed. Although economic variables might affect other decisions, the law did not allow these variables to affect the listing of species. In Spinks's words, these and other newly imposed requirements raised "serious questions of legitimate policy decisions being precluded, circumvented, or subordinated by pseudo-legalistic ploys being used as excuses for delay" in the listing process.[37] Some support for his position came from the GAO. It found that the Office of the Solicitor had approved only eight of thirty-eight DOES for proposed listings that the OES had prepared in 1981.[38] The solicitor's office claimed that the OES documents were deficient and did not provide the substantive information that the president's executive order required.

Whether economic variables were a consideration in the listing process soon became irrelevant. With the passage of the Endangered Species Act Amendments of 1982, which became law on 13 October 1982, Congress specified that future listings be based "solely" on the best scientific and commercial data available. As the Senate and House explained in their conference report on the legislation, "economic considerations have no relevance regarding the status of species and the economic analysis requirements of [the president's executive order], and such statutes as the Regulatory Flexibility Act and the Paperwork Reduction Act, will not apply to any phase of the listing process."[39] To further separate economic considerations from the listing process, Congress also changed its previous policy that had linked listing with the designation of critical habitats. The 1978 amendments specified that when a species was proposed for listing, its critical habitat should also be proposed, "to the maximum extent prudent." Such habitats could not be designated, however, until after considering the economic and other relevant impacts of the proposed habitat. The 1982 amendments retained this requirement, but designation of a critical habitat is no longer required at the time of listing. Nonetheless, the expectation is that such habitats will be designated to the "extent prudent and determinable."

These amendments also addressed at least one other concern about the listing process. The extant law allowed two years between the time a species was proposed for listing and publication of a final rule. The deadline was changed to one year, except in instances in which there is "substantial disagreement regarding the sufficiency or accuracy of the available data" relevant to the proposal. When such disagreement exists, a six-month extension is permitted. At the end of the extension, a proposal must either be finalized or withdrawn.

Responding to another congressional mandate, the Reagan administration also developed a formal ranking system to assist in the listing of candidate species.[40] This meant still another change in priority systems. When the revised system was announced in mid-1981, many biologists and environmental groups were aghast. In listing additional species, the administration proposed that the degree of threat remain the most important consideration. That, however, was not what caused the biologists' heartburn. All prior schemes (and the 1973 Endangered Species Act) had made no distinction among different classes, all were to receive evenhanded treatment. In contrast, the new system explicitly favored the listing of "higher" life forms, first mammals, then birds, fishes, reptiles, amphibians, and vascular plants over lower life forms, such as insects, mollusks, other invertebrates, and nonvascular plants. Indeed, the FWS indicated that it was unlikely to list any of these lower life forms because they would have such a low priority under the new listing system.

In justifying the new system, the FWS argued that attention to higher life forms "could allow greater concentration of resources and effort on conspicuous vertebrates, especially large mammals and birds, whose conservation probably has the greatest public support." The FWS further defended the change by claiming that by "first protecting organisms toward the top of the food chain . . . indirect protection is extended to other nonlisted components of such ecosystems."[41] It was not explicitly acknowledged, but the new system also reflected the Service's greater familiarity and longer experience with vertebrate species. A striking enunciation of this view came from the FWS itself when it once noted that it suffered "from a general lack of historic interest in the conservation of invertebrates and all but the most conspicuous plant species."[42] A related perspective came from O. Ray Stanton, who resigned as chief economist for the OES in early 1984. It was his experience, he observed, that FWS administrators approached the listing process "as an impediment to the other work of the service or as a backwater to which they have been relegated." "In some cases," he added, "an apparent conflict of interest results in less desire to list species that might prove difficult to manage, might complicate the existing management of wildlife refuges, might conflict with predator control programs . . . or might be in conflict with the desires of the service's traditional public constituency."[43]

At best, the scientific justification for the new scheme was biologically suspect, and the FWS soon conceded as much. In an internal, unpublished review of the system, the FWS emphasized that there "is danger of such an approach leading to a 'zoo' syndrome, in which attractive species are maintained through intensive management, while ecosystem vitality is largely ignored."[44]

In spite of its deficiencies, the revised system was informally used

Table 6.4   Priorities for Listing Species, Post-1983

Priority	Magnitude	Immediacy	Taxonomy
	Threat		
1	High	Imminent	Monotypic genus
2	High	Imminent	Species
3	High	Imminent	Subspecies
4	High	Nonimminent	Monotypic genus
5	High	Nonimminent	Species
6	High	Nonimminent	Subspecies
7	Moderate to Low	Imminent	Monotypic genus
8	Moderate to Low	Imminent	Species
9	Moderate to Low	Imminent	Subspecies
10	Moderate to Low	Nonimminent	Monotypic genus
11	Moderate to Low	Nonimminent	Species
12	Moderate to Low	Nonimminent	Subspecies

Source: *FR* 48 (21 Sept. 1983): 43103.

for nearly a year. It was never formally proposed, primarily because Congress deemed it unacceptable to give preference to higher life forms.[45] The FWS eventually proposed and adopted the scheme shown in table 6.4. This system retains the degree of threat as the primary consideration, but then adds two other variables. The immediacy of the threat is "intended to assure that species facing actual, identifiable, threats are given priority over those for which threats are only potential." The last consideration is intended to focus listing resources on species that represent "highly distinctive or isolated gene pools." To the FWS, this criterion has biological justification: "The more isolated or distinctive the gene pool, the greater contribution its conservation is likely to make to the maintenance of ecosystem diversity."[46] Furthermore, the loss of a monotypic genus (a genus with a single species) risks the loss of unique genetic material.

The current priority system has affected decisions about listing. The Reagan administration focused its listing efforts on native plants, birds, fishes, and mammals. The so-called lower life forms have received much less attention. During the two Reagan administrations, the FWS proposed and listed only seven crustaceans, eight insects, and nine clams. The listing of these lower life forms represented only about 5 percent of the Reagan administration's total listing activity.

These numbers become more meaningful when put into context. However unrealistic its estimates might have been in the mid-1970s about the number of species likely to be listed, by 1980 the Service had a far better idea of which species to list. More important, the Service also knows the species for which sufficient scientific information is available to justify their listing. The FWS thus was able to publish a comprehensive list in late 1980 of nearly four thousand plants that had come to its attention either through the efforts of the Smithsonian Institution or though other sources. For over eighteen hundred of these plants, the Service said that it had "sufficient information . . . to support the biological appropriateness of their being listed as Endangered or Threatened."[47] For another twelve hundred plant species, the FWS reported that it had data to support "the probable appropriateness of listing." These twelve hundred represent likely candidates for listing, depending on the outcome of further research. After additional research (and some listings), the FWS published a list of vertebrate candidates in 1982 and a revised list of plant candidates in 1983.[48] More lists of candidate species were published in 1989 and 1990, and the magnitude of the task is well reflected in the data in table 6.5. These data indicate the number of *native* species that are candidates for listing based on information already available.

When one notes the pace of previous listing activity, it is obvious that the FWS confronts an enormous task. Between the mid-1970s and the end of 1989, for example, about twenty native species were classified as endangered or threatened each year. At this pace, nearly fifty years would be required just to list species now in need of immediate protection. Even at an accelerated rate many years would pass before all the vulnerable species could receive the benefits of listing. Thousands of additional species, mostly plants and insects, are potential candidates for listing. The data also reveal some likely consequences of delay. The FWS estimates that almost two hundred candidate species may already be extinct. For another 175 species, nearly 40 percent of which are insects, the FWS once considered them for listing but it now has persuasive evidence that they are extinct. In fact, several species became extinct while the FWS was considering whether they should be listed.

Additional resources could hasten the listing process, but these resources were not available in the 1980s. In a budget submitted to Congress just before he left office, President Carter requested $3.44

Table 6.5   Candidates for Listing as Endangered or Threatened, 1989-1990

Taxa	Category[a] 1	2	3A
Mammals	7	202	9
Birds	5	54	4
Reptiles	2	53	3
Amphibians	4	50	2
Fishes	15	118	13
Snails	27	161	2
Clams and Mussels	2	59	11
Crustaceans	2	91	1
Insects	9	768	68
Sponges/Hydroids	0	6	0
Worms	0	7	2
Vascular Plants[b]	527	1,572	94
Totals	600	3,141	209

Source: Compiled from data in *FR* 55 (21 Feb. 1990): 6184-6229 (vascular plants); *FR* 54 (6 Jan. 1989): 554–79 (vertebrates and invertebrates); and *FR* 54 (19 Aug. 1989): 833 (corrections to vertebrate and invertebrate list). For a discussion of these lists, see *ESTB* 14 (Jan.–Feb. 1989): 4–5.
[a]Category 1: Species for which the FWS has substantial data on biological vulnerability and threat to support a proposal to list as either endangered or threatened.
Category 2: The available information indicates that a proposal to list is possibly appropriate, "but for which conclusive data on biological vulnerability and threats are not currently available to support proposed rules."
Category 3A: Species once considered for listing, but "persuasive evidence" now exists that the species are extinct.
[b]Includes 87 high priority taxa that are possibly extinct and five taxa believed to be extinct in the wild but known to exist in cultivation.

million for all FWS listing-related activities for fiscal year 1982. Six weeks later President Reagan reduced this request by more than 43 percent to $1.96 million. A request for a further 18 percent reduction in funds for listing came in September 1981. In spite of the cuts in resources for listing, Interior maintained that the reduced appropriations would be adequate for FWS needs.[49] This is a subjective judgment that depends on how one views what needs to be done. In its last budget proposal, for example, the Carter administration estimated that the FWS would initiate status surveys in fiscal year 1982 for approximately five hundred species to determine whether they should be listed. In the first budget the Reagan administration proposed for the FWS, the number of anticipated status surveys for the same time period had been reduced to twelve.[50]

Congress was slightly more generous than the president, but the FWS received less than $2 million for fiscal year 1982, or about half of what it had received for listing the previous year. Reagan's requests and congressional appropriations did increase in subsequent years, but by fiscal year 1987 the money provided for listing was still less than what had been available six years earlier. Accounting for inflation, and considering the additional efforts required to list a species, the reduction in funds was especially harmful to the FWS.

Uncertainty over additional congressional authorizations in the late 1980s also frustrated efforts to plan effectively. The 1982 amendments to the Endangered Species Act provided authorization for funding the FWS's implementation of the act for three fiscal years (from October 1982 through September 1985). If past practices had been followed, Congress would have passed new legislation (before October 1985) authorizing additional expenditures for a multiyear period. This did not happen. The House approved a three-year extension and additional funds in 1985. In the Senate, however, the proposed listing of the Concho water snake, which is found only in Texas, led Senator Lloyd Bentsen (D.–Texas) to stall action. Bentsen, trying to protect the construction of a dam, said that concern for the snake represented a "serious case of misplaced priorities." Due to Bentsen's delaying tactics, the Senate recessed before a vote on the three-year extension could take place, so Congress continued to fund the FWS at existing levels. Efforts to pass a comprehensive, multiyear reauthorization succeeded in 1988, but only after failures in the two previous years.

Conclusions

The previous chapter began with a discussion of decision rules and how they might affect the listing of endangered species. Such rules can provide broad discretion and facilitate conscientious implementation. Decision rules can, in contrast, severely constrain administrative agencies and their overall effectiveness. A review of the listing process reveals that these rules have changed enormously. The secretary of the interior was first granted authority to list endangered species in 1966. All that was necessary to list a species was a determination, after consultation with the affected states, that a species' existence was endangered. The burden of proof was on Interior, but this burden was a relatively light one. There were no requirements for public participation or the publication of scientific evidence, and no deadlines were applicable. Scores of native species were listed relatively inexpensively and without major controversy.

Much has since changed. After more than two decades' experience, the listing process is costly, complex, cumbersome, and time consuming. Before a species can be listed, all of the following steps are normally necessary:

1. Develop and use a priority system to determine which species should be listed;
2. Evaluate available data and review the species' status;
3. Consider existing protection efforts;
4. Determine whether to list the species as endangered or threatened;
5. Determine whether critical habitat should be designated and, if so, conduct an assessment of the likely economic impacts of designation;
6. Prepare a proposed rule that summarizes the threats the species faces and discusses the conservation actions that will result from listing;
7. Publish a proposal in the *Federal Register* that solicits public comment;
8. Notify appropriate federal, state, and county agencies and invite comments from them;
9. Give notice to appropriate professional scientific organizations;
10. Publish a summary of the proposed regulation in newspapers of general circulation in the area where the species is believed to exist;
11. Hold at least one public hearing if so requested;
12. After receiving comments, determine whether substantial disagreement exists regarding the sufficiency or accuracy of the available data;
13. Ensure that deadlines are not exceeded;
14. Receive appropriate administrative approvals from within the

Division of Endangered Species and Habitat Conservation, the FWS, and the Department of the Interior;

15. Either withdraw the proposal or prepare and publish a final rule that discusses all comments from whatever source; and,

16. Wait thirty days after publication in the *Federal Register* for the listing to become effective.[51]

Still other administrative steps and procedures are involved with foreign species and with species under consideration for listing because of petitions.

In addition to the legislative requirements and internal FWS procedures, other factors that affect the listing process include, at various times, politics, public opinion, budgetary constraints, economic impacts, the relative status of the OES within the FWS, and sensitivity to congressional concerns and to the projects of other federal agencies, to name only a few of the most important.

Congress has further circumscribed the listing process by stipulating that *anyone* can sue Interior when it fails to perform duties—such as the listing of species—that the Endangered Species Act mandates. Administrative agencies have a natural aversion to legal action, and one way to avoid such action is to steer clear of controversial decisions. Few listing actions have ever been challenged in the courts, but the possibility of a lawsuit has a noticeable deterrent effect on FWS activity.[52] Few administrative agencies, including the most powerful, remain unscathed by cases of the magnitude of *TVA* v. *Hill*, the snail darter case.

In short, the changes since 1966 have altered the burden-of-proof requirements associated with the listing process. For most species the locus of decisionmaking remains the same, but the evidentiary requirements have multiplied. Decisions are far more technically and procedurally elegant than they were in the past. In contrast, the additional listing requirements probably reduce the likelihood that many vulnerable species will receive the benefits of any doubt. When doubt exists or when sufficient information is unavailable, the existing listing requirements encourage delay or additional research. Having "sufficient" information is always desirable, but there never has been enough money to ensure that the requisite information will be available for all candidate species. As a consequence, a shortage of resources and the current listing procedures and their implementation will doom some species to extinction. Moreover, waiting for conclusive evidence will mean that most species are listed only when they are at the brink of extinction or when it is too late to employ effective solutions for their plight.

At least one alternative exists to the present arrangement for listing species. Currently, no species is officially designated as endangered or threatened until after many procedural barriers have been overcome. The alternative, though surely controversial, would give vulnerable species the benefit of doubt, conserve valuable resources, and alter the burden of proof. Quite simply, lack of adequate information about a species' status would be regarded as evidence of endangerment until such time as someone could prove otherwise. This might mean, for example, that all species in categories 1 and 2 (see table 6.5) would be listed without the agency having to comply with existing procedural requirements. The species would remain classified as either endangered or threatened unless information was provided to the contrary. No doubt some mistakes would be made, but the mistakes would contribute to the species' benefit, not to their detriment or possible extinction, as is now often the case.

However important the listing process, it is only the first stage in protecting endangered species. There are benefits related solely to listing, but the threat to many species is an indirect one. For many, their continued existence is in doubt because their habitats are altered, degraded, or destroyed. The next chapter examines how and how well the habitats of endangered species are protected.

T he listing of species can be an important step in their protection because certain restrictions exist in regard to the taking of these species. Despite the potential benefits of listing, it serves primarily to identify species in need. For many listed species, the need is for viable habitats.

Species survive in habitats that are smaller than their historic ranges, but a decline in a habitat's size almost always leads to a more precipitous decline in the value of what is left. As a rule of thumb, notes Norman Myers, "arithmetic loss of space leads to geometric decline in the value of the remaining space." Other research on the effects of diminished habitats finds that extinction rates increase as habitat size decreases.[1] This pattern is evident even in areas free of widespread development. In a study of fourteen national parks in western Canada and the United States, as an example, William D. Newmark found that all but one park had experienced the loss of several mammalian species since the parks' establishment.[2] These mammals were not extinct outside the parks, but they had not been sighted within the parks in the previous ten years. The smaller the park, the higher the proportion of species no longer sighted within it. The two smallest parks in the study had both lost over 20 percent of their mammalian species since being established. In contrast, the largest park, which is over 8,000 square miles, has not experienced any decline in the number of mammalian species since its creation in the early 1900s.

In only a few instances did a species disappear from a park because of specific human intervention, such as excessive hunting. Indeed, over 85 percent of the species disappeared because of natural losses. The parks were too small to sustain the populations, which became isolated in what one scientist has called a "sea of man-made habitat." The parks provided some safety and security, but the lost species were not immune to the adjacent disturbances and developmental activities or, apparently, to the milder disturbances within the parks. These disturbances continue and, since species in an ecosystem are mutually dependent, additional losses of mammals are likely as are losses of birds and plants, which Newmark did not assess.

If loss of habitat leads to a decline in biological diversity, does a subsequent increase in habitat size improve the prospects for recovery? Some researchers speculate that a tenfold increase in a protected area's size will about double the number of species whose chances for long-term survival are improved.[3] Providing additional habitat may not be as easy as one might hope. The problem is not limited to just providing more acreage; unless this acreage offers a habitat suitable for a species' existence, the species will not survive in it. Moreover, efforts to provide suitable habitat must consider the political ramifications if humans must be inconvenienced in order to provide this habitat.

People's willingness to change their locale or behavior to benefit species is likely to be a function of at least two factors—economics and the nature of the species. The Supreme Court has ensured that economic factors will be a consideration in many instances in which local governments restrict landowners' use of their property. In *First English Evangelical Lutheran Church* v. *County of Los Angeles*, the Court ruled that when governments deprive owners of the use of their property, even temporarily, they must be compensated.[4]

As public opinion polls reveal, a hierarchy of attitudes exists toward endangered species. Some are viewed much more (un)favorably than others, even among officials charged with protecting the species.[5] Attempts to relocate people or change their life-styles on behalf of a threatened snake or an endangered kangaroo rat are not likely to be well received. When people complain to elected officials and ask if they favor rodents and reptiles over incensed voters, responses are easily predicted. Agencies that foolishly propose such "dubious" schemes should expect to find themselves excoriated before congressional committees and threatened with budget cuts and perhaps even major changes in the laws governing their activities.

In contrast, the consequences of not providing suitable habitats are equally predictable. Some scientists believe that if appropriate habitats are not assured, then a "veritable rush of extinctions" will follow.[6] This conclusion further suggests that most vulnerable species need an area of some critical minimum size if they are to survive. The problem is in the determination of habitat size and the identification of methods of protection.

Ascertaining an Appropriately Sized Habitat to Protect

Destruction or alteration of habitats is the major threat to most endangered species. One solution to such alteration is to reduce or eliminate the threat. This is exactly what the Public Land Law Review Commission (PLLRC) recommended in its 1970 report to the president and Congress. "Where certain areas of public lands are the only or best habitat of species that may be threatened with extinction," said the commission report, "other uses of the land and resources should be foregone or restricted in the interest of protecting them."[7] There is simplicity in this idea, but this solution is both infeasible and politically unacceptable. The United States prides itself as a nation that fosters growth and that largely respects landowners' decisions about private land use. Moreover, zoning laws are rarely found in the rural areas where endangered species are most likely to be found. Public lands under the control of the USFS and the BLM are managed for multiple usage in order to accommodate a wide variety of interests. Attempts to establish a national strategy to preserve open space have met repeated failure. Equally important, restricting access to certain areas would not guarantee the survival of all vulnerable species.

If all further adverse modification of habitats were halted, existing developmental activities might still imperil thousands of species. These would include many species already listed plus the thousands of candidate species now awaiting listing. Current consequences of past habitat alterations might soon put still other species in jeopardy.

This inability to eliminate the threat to habitats creates an unfortunate paradox. What needs to be done—the preservation and protection of ecological communities—usually cannot be done to the extent necessary. Areas that are set aside for a species' benefit often tend to be too small. Increasing their size is a possibility, but the benefits (i.e., the number of species protected) may not match the political or economic costs. In addition, specialized habitats are rarely recoverable once humans have placed their imprint on them. Cleared forests, drained swamps, and dammed rivers are unlikely to be returned to their predevelopment state, regardless of the effort or good intentions of those involved.

The consequence is that an effective means of preservation, namely eliminating the causes of habitat degradation, is a viable solution for only a few species. What is left, then, is to preserve, to the extent possible, the habitat of species that suffer because of intrusions into these habitats. In order to do so, several questions have to be answered. What is the species' habitat or range? How large an area should be protected? How should the habitat be protected? Related questions focus on the priorities for and decision rules applicable to designating these habitats. Which habitats should be protected first? What procedural steps are involved with designation? Who has the burden of proof? As will be seen, scientists can provide information relevant to many of these questions, but nonscientific considerations affect the ultimate choices—just as is the case for listing species.

What is the Species' Habitat or Range?

A species' habitat includes its native environment and that area necessary for its life, survival, and growth. For locally endemic species, especially plants, the task of identifying habitat is a relatively easy one; by definition, their range is small and localized. Many species cannot migrate because of their location on islands or because of their dependence on a highly specialized environment. For other species, identifying habitats is far more difficult. Large mammals might forage in an area of several hundred square miles over a year's time. Some migratory bird species travel thousands of miles annually, with many stops along the way. Whooping cranes migrate approximately 2,500 miles from northern Alberta and the Northwest Territories to Aransas National Wildlife Refuge, near Corpus Christi, Texas. Though their annual migratory path remains relatively stable, the endangered cranes' stopover sites vary.

A further problem relates to the dynamic nature of habitats. As a species experiences periodic fluctuations in its population, the species' range and/or habitat might also contract or expand. A habitat can also depend on water flows, seasonal variations, the availability of food, and other, unknown or poorly understood ecological considerations. At best, for many endangered species, it is difficult to determine with certainty what within their range or habitat renders it suitable for their survival. In turn, this uncertainty creates problems for those who justifiably seek certainty.

How Large an Area Should be Protected?

Once a present habitat is identified, decisions must be made about how large an area will be set aside for the species' benefit. How this issue is addressed provides an important indicator of commitment to the protection of endangered species. An historic range could be designated as the habitat that should be protected, but for many species such a designation would be futile. Areas would be identified although no chance exists that a species would ever repopulate them. This does not make political sense. For most endangered species much of their historic range has been so altered that any attempt at repopulation will fail. In other instances, the historic range would cover such a large area that designation would be meaningless. Wolves and mountain lions once roamed much of the United States but are now found in only a few isolated areas. By way of contrast, many species survive and frequently thrive in areas much smaller than their historic ranges.

Another alternative would identify the present habitat and then designate a buffer zone to provide room for expansion or anticipated recovery. This approach has some merit because the goal of all endangered species legislation is to bring about the recovery of species in jeopardy, not just the survival of their remaining members. For many species their endangerment means that something is adversely affecting their habitat or that its carrying capacity is less than is needed. Protecting the present habitat and then providing some additional area might address either or both of these concerns.

This alternative is not without problems. On the one hand, scientists may not know or be in a position to justify the appropriate amount or location of additional space. As with the first alternative, scientists may be able to offer fairly accurate estimates. These will not, however, be adequate for those expected to provide certainty to landowners, developers, or others whose activities might be affected. They will want readily identifiable legal descriptions that scientists will view as no more than imprecise estimates of a species' habitat.

On the other hand, protecting the added area presumes that the affected species will find the area to be a hospitable one. Although it may at one time have been part of the historic range, much may have changed so as to make the area entirely inhospitable. Furthermore, the mere hope that a species will expand its range is not likely to convince policymakers to deter developments with economic appeal.

The last choice is the most politically astute, but it is likely to be the riskiest, at least from the species' perspective. This alternative identifies and protects the present habitat of listed species. This approach is based on several important assumptions. First, in delineating the present habitat, one must assume that it is sufficiently large to meet the species' needs. For species that are already in jeopardy because of a threat to their habitats, this is problematic. Second, because recovery is a goal, an assumption must also be made that the designated habitat (and the protection afforded it) will be adequate to sustain a larger population and to guarantee the survival of the species. Merely increasing a species' population is unsatisfactory if the larger number exceeds the carrying capacity of the habitat. Third, protection of the present range or habitat may be adequate but only when there are assurances that nothing outside the habitat alters the essential features within it. The spraying of pesticides, withdrawal of water for irrigation, or soil erosion or siltation can affect habitats many miles away. In these instances, protecting only a formally designated habitat might well be useless.

A hybrid approach would select a habitat to be protected on the basis of a species' characteristics, such as prospects for recovery or recolonization of previous habitats, the potential effects of anticipated developments, or perhaps economic or symbolic value. Both politicians and the public are more likely to be tolerant of the needs of the bald eagle than of the wartyback pearly mussel. The hybrid approach treats each species in an individual manner, and this makes ecological sense. From a bureaucrat's perspective, however, a species-by-species approach is undesirable because it prevents the use of "safe" standard operating procedures. It is more difficult to justify unique choices than it is to claim that a decision was made in conformance with long-standing precedents and existing agency guidelines.

For each of the alternatives, some irony exists. The approaches that are the least politically acceptable are also the ones that incorporate the greatest optimism about a species' recovery. As political acceptability increases, risk to a species might also increase.

How Should Habitats be Protected?

To allow a species to recover from its endangerment means, at the least, removing the threat to the species. When this threat comes in the form of altered or destroyed habitats, decisions must be made about which activities, if any, will be limited, regulated, or prohibited in or adjacent to the habitat. Not all human activities will be harmful to the species and, in some instances, they can even be beneficial. To improve the chances of recovery some controls must be imposed on those activities that disrupt the habitat; the difficulty is in determining detrimental intrusions before their occurrence.

One might suggest restrictions on the taking of an endangered species.[8] There can be situations, however, when some taking benefits a species, biologically and perhaps even politically. Taking might be justified when it is for research, captive propagation, or for thinning a population that is exceeding the carrying capacity of its habitat, an unusual possibility. Eliminating some individuals from a habitat they are overtaxing might prevent a catastrophic population crash. From a political perspective, the controlled killing of some predators, such as wolves that attack livestock, might limit criticism from elected officials and their constituents or forestall an erosion in support for the protection of other endangered species.

In discussing each of the three major questions it is apparent that straightforward answers are not available. Scientists rarely have the information they need about particular habitats, yet the need for action is evident. Inaction encourages extinction, yet whatever action is recommended must incorporate concern for the political environment in which decisions are ultimately made. A further problem relates to decision rules and priorities. Do the rules facilitate protection of habitat and, given the large number of species in trouble, on what basis are decisions made to protect some habitats at the expense of others?

Cautious Beginnings

The Endangered Species Preservation Act of 1966 had little to say about the protection of habitats. Protection was not one of the law's stated purposes. The law recognized modification and destruction of habitat as causes of endangerment, but it did not obligate federal agencies to alter their activities to protect habitats. Instead, federal agencies were expected to preserve the habitats of endangered species on lands under their jurisdiction insofar as was "practicable and consistent with [their] primary purposes."[9] In other words, federal agencies did not have to subordinate their primary missions to the protection of endangered species or their habitats. A species' habitat might be in jeopardy, but the federal agency causing the threat could exercise discretion in responding to it. If protection could be accomplished *and* it did not conflict with an agency's primary goals, the habitat might be protected.

The chance of this occurrence was not always good. No incentives were available to encourage a change in agency practices, and several things probably discouraged effective efforts to preserve habitats. The secretary of the interior was authorized to utilize programs within his department to achieve the act's goals. For other agencies, however, the secretary was instructed to encourage them "to utilize, where practicable, their authorities" to protect species. Few agencies would want to acquire the biological competence needed to assess the consequences of their activities on habitats. The qualified nature of the law's care-for-habitat provision also signaled federal agencies that Congress considered their primary missions to be of greater importance than the preservation of biological diversity. Agencies are never enthusiastic about having to spend their limited resources on programs that do not contribute to their own well-being. For all these reasons, protection of habitats was not likely to be a high priority, even within the Interior Department.

Subsequent legislation, passed in 1969, did little to change the thrust of the meager efforts to protect habitats. The Endangered Species Conservation Act of 1969 did, however, continue the previous act's authorization for the secretary of the interior to purchase habitat through the LWCF Act of 1965. As noted in chapter 5, this law authorized Interior to spend up to $15 million to acquire the lands or waters needed to conserve, protect, or propagate fish and wildlife threatened with extinction.[10] No more than $5 million could be spent in any one year and, unless Congress approved, no more than $750,000 could be used for any single area. The 1969 law made only minor modifications in these provisions. The single-area limit was also raised to $2.5 million in 1969, and Congress authorized the expenditure of an additional $1 million per year (for three years) to acquire privately owned land within areas the Interior Department already administered.

Using appropriations from the LWCF, Interior purchased over 35,000 acres of habitat for ten endangered species between October 1966, when the Endangered Species Preservation Act was passed, and June 1973.[11] The location of this acreage gives some idea of Interior's priorities for habitat preservation. First, the degree of threat facing all species listed as endangered did not seem to be a determinant of which habitats were purchased. In early 1970, as an example, Interior identified twenty-seven birds and mammals as those in the greatest danger of extinction.[12] Of the twenty-seven, Interior purchased habitat for only two, the southern bald eagle and the Columbian white-tailed deer. Consequently, only 15 percent of the acreage acquired through fiscal year 1973 and only 45 percent of the funds spent went for species in the greatest danger of extinction. In contrast, over 40 percent of the money spent in the first six years was used to buy land for *sub*species eventually determined to face a low degree of threat *and* a high recovery potential.[13]

Second, although thousands of acres were purchased, nearly 60 percent of the total was for alligators in the Okefenokee Swamp in southeastern Georgia. The alligators there were not in particular danger, but the land was cheap, averaging about $40 per acre. At over $2,500 per acre, habitat purchased for the bald eagle was the most expensive.

Even in the face of this expense, Interior did not spend all the money initially authorized. As early as 1971, the department had asserted that $15 million would not be enough to purchase all the habitat that required protection. At the least, the department said, an additional $24 million would be needed to complete its planned acquisitions of habitat. Having said this, the department soon realized that it could not acquire habitat as quickly as it would like. By mid-1973, it had spent slightly more than $11 million for habitats. Indeed, so much unspent money remained available that Congress did not provide any money for habitat acquisition for endangered species for the fiscal year beginning in July 1973.

Recognition of the Importance of Habitat

The 1966 and 1969 laws reflected a lack of assertiveness in the protection of habitats, but this changed with the passage of the Endangered Species Act of 1973. Section 7 of the new law required federal agencies, including Interior, to be far more conscientious in their protection of habitats. As the House report on the 1973 law noted, many things threatened species' continued existence, but the most significant of these is the destruction of critical habitat. In recognition of this concern, one purpose of the new law was to "provide a means whereby the ecosystems upon which endangered species and threatened species depend may be conserved. . . ."[14] This statement provided the first statutory recognition of the importance of preserving ecosystems rather than just species.

To address the concern for habitat, Congress removed the limit on the total amount of money from the LWCF that could be used to purchase habitats. The new law also authorized the purchase of habitats for plants, which had not been eligible for listing or habitat protection prior to 1973.

Far more important, the 1973 law stipulated that all federal agencies would henceforth be *required* to ensure that their activities do not jeopardize the continued existence of listed species or result in the destruction or modification of habitats that are considered to be critical to listed species. To ensure compliance, all federal agencies are required to consult with the secretary of the interior about how their projects would affect listed species and their designated critical habitats. In the words of the law: "All other Federal departments and agencies shall, in consultation with and with the assistance of the Secretary, utilize their authorities in furtherance of the purposes of this Act."[15] These purposes included the protection of habitats and the conservation of listed species. Conservation, declared Congress, means "the use of all methods and procedures which are necessary to bring any endangered . . . or threatened species to the point at which the measures provided pursuant to this Act are no longer necessary."[16] In short, the 1973 law represented a fundamental change in existing policies and congressional expectations.

However well intentioned the change, the provisions in Section 7 provided almost a certain guarantee of contentious implementation. Understandable statutes facilitate effective implementation, but the absence of clarity often produces just the opposite result. In drafting the new law, Congress seemed to prefer the latter. The Endangered Species Act explicitly stated its goals, but the law was much less precise in specifying the means to achieve these goals. The law did not define critical habitats. With the exception of a requirement that affected states be consulted, the law also failed to indicate what procedures should be used for the designation of critical habitats or list the kinds of information that would be required to justify their designation. The law was similarly vague about the consultation requirement. Consultation would be mandatory, but the expected content of the consultation was left unsettled as were: a) the obligations of the agencies involved in the process; b) whether there would be a limit on the length of the consultation process; c) the means of enforcing the consultation requirement; and, d) the sanctions for noncompliance.

Another indication of problematic implementation with Section 7 involved the amount of change that the new law mandated. Successful policy implementation often depends on the level of consensus and the amount of change required.[17] Policies characterized by high consensus and slight change are the best candidates for success. Conversely, failure or ineffective implementation is likely to be a product of low consensus and the need for large change.

The Endangered Species Act of 1973 called for fundamental changes. With the two earlier endangered species laws, the primary missions of federal agencies could take precedence over the protection of endangered species and their habitats. The Endangered Species Act was intended to reverse this order.[18] A federal agency could no longer proceed with its activities unless it had first considered the impact of these activities on listed species and their habitats. In the future, agency missions would be secondary and subordinate to the protection of listed species and their habitats, or so Congress seemed to intend when it imposed these new obligations.

Few governmental agencies actively seek additional responsibilities; most already have enough to do. An agency occasionally may be willing to assume new functions, but these must be ones that provide some advantage, such as additional resources, jurisdiction, or favorable publicity. Likewise, an agency rarely responds favorably to unfamiliar duties imposed on it, especially when these duties are viewed as relatively unimportant to its primary mission.

Experience with the National Environmental Policy Act (NEPA) offers an instructive lesson in agencies' willingness to incorporate environmental values into their decision-making processes. The NEPA obligates agencies to prepare environmental impact statements for all "major Federal actions significantly affecting the quality of the human environment."[19] Though widely heralded as a panacea for environmental problems, the statement is only advisory. When damage to the environment is anticipated, the project agency is not required to avoid or mitigate the damage. Equally important, the NEPA allows a balancing of environmental, economic, and other values.

Serge Taylor has offered an interesting assessment of the NEPA process and, by extension, a useful comparison with Section 7 of the Endangered Species Act. For the former, Taylor believes that it substitutes analysis for reorganization. In his words, since the sponsors of the NEPA "lacked sufficient power to change the decision premises of all agencies directly, they tried to change agency policies indirectly by requiring a different type of information to enter the decision-making process."[20]

Regardless of the legislative strength of the advocates of the Endangered Species Act, their goal was to change decisionmaking premises, not just the kind of information that agencies considered. What this means then is that compared to Section 7, the NEPA imposes rather modest and flexible demands on federal agencies. Yet the bureaucratic reaction to the NEPA was anything but favorable. Walter Rosenbaum offers a revealing illustration of how agencies initially responded to the need for environmental impact statements:

> Administrators long accustomed to the role of advocate for their particular agency projects were now expected to engage in self-criticism and searching examination of the ecological value of their efforts. In the beginning, ambiguity and confusion reigned. The specific procedures that agencies had to follow to produce a satisfactory impact statement were unclear. The meaning of critical words and phrases in the law had to be interpreted through administrative and judicial rulings. Many agencies lacked the technical resources to prepare the required statements.[21]

Many agencies were reluctant to comply with the requirements of the NEPA. Some officials believed that their projects were exempt from the requirement that impact statements be prepared. Other agencies resisted the expense of preparing the statements, and still others complained about delays in their projects because of the time needed to prepare statements. Within four years of the law's passage, over four hundred law suits had challenged federal agencies' implementation of the law.

If the modest and flexible requirements of the NEPA caused such consternation, what would be the effect of a law that did not allow any balancing or consideration of costs and benefits and that *required all agencies to use all measures necessary* to protect listed species? Here again interesting parallels exist with the NEPA. Some people initially believed that the language of the NEPA was intended to be symbolic and that Congress did not fully consider the implications of requiring federal agencies to incorporate environmental values into their decisionmaking. Much the same can be said about congressional consideration of the single section of the Endangered Species Act requiring consultation and the no-jeopardy standard. The legislative history of the pertinent section is relatively brief. The House report on the proposed law provided only the most limited direction on how the provision should be interpreted or implemented, and the conference report on the act omitted entirely any meaningful discussion or explanation of Section 7.[22] Several interpretations are available for this seeming neglect. The virtual absence of a legislative history may have been intentional.[23] Recognizing how consequential the requirements of Section 7 might be, proponents supposedly understated their significance. If this is the case, the subsequent history of these requirements makes the proponents' accomplishments all the more remarkable. One participant in the implementation process would eventually declare that Section 7 includes "the two most controversial sentences in the history of the conservation movement."[24] Examination of how the FWS implemented the section allows some explanation for this judgment.

Developing Regulations to Implement Section 7

The Endangered Species Act became effective once President Nixon signed it at the end of 1973. The consultation and no-jeopardy requirements were thus applicable to all federal agencies, and the FWS could immediately begin the formal designation of critical habitats. Given Congress's failure to provide any procedural details, developing regulations to implement Section 7 would be a lengthy and contentious process.

The FWS had at least two contrasting choices. It could aggressively designate critical habitats and stringently enforce the consultation and no-jeopardy provisions. In doing so, the FWS would be responding to the idealistic expectations embodied in the law and would find itself with a unique opportunity to expand its bureaucratic influence and prominence, but only at the expense of other agencies. Alternatively, being too assertive could be counterproductive. Administrative agencies succeed when they maintain friendly relations with other agencies. In contrast, they increase their chances of failure when they alienate those from whom they expect cooperation and compliance.[25] Regulatory agencies often find themselves at a disadvantage in this contest for allegiance. In enforcing Section 7, the FWS would be implementing regulations potentially influencing most federal agencies. With so many factors affecting how these agencies would respond to the section's mandates, the FWS might be inclined to display a natural bureaucratic tendency to avoid uncertainty.[26] This would require the FWS to negotiate with its political environment and to ensure that Section 7 would neither offend its sister agencies nor place the FWS in a situation in which it might find itself accused of meddling in the affairs of its bureaucratic colleagues.

As some observers of the regulatory process have emphasized, when a regulatory agency wants to be successful, "a tendency to bargain for compliance and cooperation may generally be wise policy, whatever its susceptibility to abuse or failure in particular areas."[27] Others have gone so far as to argue that successful regulation depends on the approval of those who are regulated.[28] This consent, so the argument goes, comes about through negotiated settlements that are acceptable to the regulatory agency and the regulated parties.

The path the FWS chose seemed to be a curious mixture of both approaches. The agency initially revealed an aggressive approach to the designation of critical habitats, but a far more timid reaction to the consultation requirement.

Fifteen months after the law's passage, the FWS offered its first indication of how it would define critical habitats. In a statement interpreting the agency's understanding of such habitats, the FWS provided a broad and expansive definition. It emphasized that a species' habitat "could be considered to consist of a spatial environment in which a species lives and all elements of that environment including, but not limited to, land and water area, physical structure and topography, flora, fauna, climate, human activity, and the quality and chemical content of soil, water, and air." The statement further indicated that in determining *critical* habitat, the following criteria could be considered:

1. Space for normal growth, movements, or territorial behavior
2. Nutritional requirements, such as food, water, and minerals
3. Sites for breeding, reproduction, or rearing of offspring

4. Cover or shelter
5. Other biological, physical, or behavioral requirements[29]

In interpreting this definition, the FWS emphasized that designated critical habitats would "not be restricted to the habitat necessary for a minimum viable population," thus suggesting that critical habitats would be sufficiently large to accommodate expansion and recovery.

Whatever critical habitat is designated, the intention is not to preclude all human activity within it. As Keith Schreiner once explained, determination of a critical habitat would not "place an iron curtain around a particular area . . . it does not create a wilderness area, inviolable sanctuary, or sealed-off refuge."[30] Furthermore, the consultation and no-jeopardy requirements apply *only* to federal agencies and actions they fund or authorize, and then only in instances in which the action "might be expected to result in a reduction in the numbers or distribution of [a species] of sufficient magnitude to place the species in further jeopardy, or restrict the potential and reasonable expansion or recovery of that species."[31] As a consequence, the designation of a critical habitat might be of greater concern to federal agencies than the listing of a species. For these agencies, then, there would be justifiable concern about how many habitats would be designated as well as their size and location. Once an area is designated as a critical habitat, Section 7 imposes an absolute obligation on federal agencies to ensure that their activities do not jeopardize a listed species' continued existence or destroy or adversely modify its habitat.

It was not long before the FWS provided justification for whatever apprehension existed among other federal agencies. Testifying before a congressional committee in 1975, Lynn Greenwalt stressed that with "a proper array of critical habitat designations across the map of the United States, we could seriously impede, if not halt, many activities carried on either by the Federal Government or under Federal sponsorship in some fashion."[32] If the FWS had its way, such an array of designations would soon be forthcoming. The agency said that when it proposed the listing of a species in the future, the Service would also propose the designation of a critical habitat when desirable or appropriate. The FWS further stated its intention to designate critical habitats on a priority basis for most native species listed before the issuance of the new FWS regulations.[33] For ten, high-priority species, critical habitat would be determined as rapidly as possible since they would most likely benefit from designation.

By the end of 1975, the FWS had proposed critical habitats for seven species, including five high-priority ones (the Indiana bat, the Florida manatee, the whooping crane, the California condor, and the Mississippi sandhill crane). Fifty-seven additional proposed designations were published in the next two years as well as final rules for twenty-two critical habitats, including six high-priority species.

The potential for controversy seemed to be an irrelevant consideration to the FWS. The first final designation of a critical habitat was for the snail darter. Other final rules for critical habitats published in 1976 included designations for the Indiana bat, the California condor, and the Florida manatee. The initial appearance was one of assertiveness on the part of the FWS, but there were several discordant signs and criticisms of its actions.

The FWS was supposed to designate critical habitats, but it was also expected to provide guidance to other federal agencies about the obligations that the Endangered Species Act imposes on them. What must these agencies do to comply with Section 7? What is expected of them in the consultation process? What actions are required to protect listed species and their critical habitats?

When Congress passes a law, the agency with lead responsibility for its implementation normally develops regulations and publishes them in the *Federal Register*. These regulations clarify the law and amplify how the lead agency intends to administer it. Once published these regulations have the force of law. So long as agencies comply with the regulations, they can reasonably expect that they will not be sued for violating either the applicable law or the related regulations.

Federal land managers might also reasonably have expected that the FWS would publish these regulations before it designated any critical habitats. From the Service's perspective, however, this might not be a rational bureaucratic strategy. Faced with the law's vagueness and Congress's inability or unwillingness to provide more explicit guidance about Section 7, the FWS was likely to be cautious in what it did, especially because of the agency's limited bureaucratic clout and second-class status within Interior. This caution manifested itself in the form of delay.

Not only did the FWS delay in issuing any regulations, but it also took a somewhat different path than might have been expected from an aggressive regulatory agency. Rather than publishing binding regulations for Section 7, the FWS developed advisory guidelines for consultation, but only after other federal agencies requested that it do so and had advised the FWS on what to include in the guidelines.[34] Moreover, the guidelines adopted a conciliatory approach reminiscent of the qualified expectations found in the 1966 and 1969 laws. The guidelines, published in April 1976, stated that their use was not mandatory, and that federal agencies could apply the guidelines at their discretion. In fact, the guidelines would not be published in the *Federal Register* because they were intended only for the "internal guidance" of these agencies.

This was noteworthy because Section 7 imposes a mandatory duty on all federal agencies. They are required to do whatever is necessary to protect listed species and their designated critical habitats. In this regard the Endangered Species Act could not have been clearer. Ironically, however, instead of stating that preservation of species and their habitats should be a preeminent consideration in agency decision making, the FWS advised federal agencies that the guidelines were not intended "to hinder, halt, or preempt actions or statutory requirements of other Federal agencies."[35] At the least, this statement implied that the primary missions of federal agencies could excuse their obligations to protect listed species and their habitats.

In addition to these weaknesses, the guidelines created further ambiguity about the meaning of jeopardy and adverse modification. The FWS appreciated the need for federal agencies to know which of their actions would jeopardize endangered species or their habitats. Instead of providing definitions, the FWS relied on what it called "working concepts" of jeopardy and adverse modification.[36] These concepts were designed to assist in the clarification of the Endangered Species Act. Without a definition, however, a federal agency might not know or believe that its actions actually jeopardized a species. This problem could be overcome if the FWS assumed the responsibility for collecting and analyzing data on other agencies' projects and then determined whether a species or its designated habitat was in jeopardy as a result. The FWS did not adopt this approach; it had neither the staff nor the legal authority to do so. The guidelines reminded agencies of their obligation to review the potential effects of their activities but then let each agency decide for itself whether its activities jeopardized an endangered species' continued existence. This procedure would work to the benefit of endangered species only if all agencies had the scientific competence to make such judgments and if these agencies could objectively appraise their own projects. The competence might be available, but administrators would have a conflict of interest. They could hardly be expected not to give their own projects the benefit of doubt.

All of these concessions revealed the Service's willingness to accept less than full compliance with the spirit of Section 7's mandates. Matthew Holden's analysis of pollution control offers a graphic description of the situation. In appearing to accept good intentions as evidence of compliance, the FWS would be "making a deal with the party whose compliance is necessary." That deal, added Holden, "asserts that if the regulated party will take just enough action to permit the regulator to ward off public pressure, then the regulator will be understanding and sympathetic about the regulated party's difficulty in not complying in the detail which would produce a drastic change in the environment." [37]

The FWS did offer its assistance and expertise to its sister agencies, but this offer contained some potential disadvantages to these agencies. If consultation revealed that jeopardy to a species or its critical habitat was a possibility, the FWS suggested its willingness to "initiate appropriate surveys, studies, research and other means to obtain data and information relative to the impacts of the activity. . . ." [38] These could be both expensive and time consuming. The latter would surely delay and possibly halt an agency's project.

The last major difficulty with the guidelines was their implicit assumption that the FWS and the consulting agency could eventually agree on a project's anticipated effects on listed species and on the appropriate remedial measures. This assumption reflected the limited congressional debate on the consultation process. In reviewing this legislative history, it is obvious that the hypothetical examples discussed in Congress were those that could be easily remedied with only minimal inconvenience to the public. [39] These could well have been atypical examples. As the number of listed species and critical habitats increased, so also would the probability of conflict and deadlock, something most members of Congress did not seem to appreciate. This probability would increase if the FWS interpreted the Endangered Species Act in the way Congress seemingly intended. With Section 7's absolute mandate against jeopardy or destruction of habitat, strict interpretation would mean that all compromise would have to come from the action agency, not the FWS. This situation would preclude the bargaining that is typical of the implementation of most other laws. In sum, the guidelines on Section 7 potentially created more problems than solutions.

Suffering the Consequences of Ambiguity

According to the FWS, designation of critical habitats is based on the scientific and ecological needs of the affected species. Just as with the listing process, this designation is expected to be scientifically defensible and free from political interference. In fact, however, the FWS soon found itself, on the one side, willing to negotiate the acceptability of critical habitats and, on the other side, accused of introducing nonscientific considerations into the designation process.

In the first instance, the FWS established a procedure whereby it would notify affected agencies prior to proposing designation of a critical habitat. This notification would request information, assistance, and recommendations, to which the FWS would attempt to be responsive.[40] For other agencies within Interior, however, the FWS accepted a different procedure that placed it in a more compromising position.

Two of the department's prodevelopment units, the BOR and the BLM, were concerned about the potential consequences of designating critical habitats on lands they administered. Interior's response to these concerns was to require the FWS to notify the two bureaus whenever it was considering a designation that would affect either bureau or any of their projects. Once notified, the bureaus could provide comments and recommendations to the FWS. This procedure was the same for all agencies. Before the FWS was allowed to propose a habitat affecting either bureau, however, the FWS had to notify the bureau and justify its intentions. Either bureau could object to the proposed designation, and the FWS was then obligated to "make every effort" to reach agreement.[41] If agreement could not be reached, the dispute would be referred to the appropriate assistant secretaries within Interior or, when necessary, to the department's secretary. Given their relative importance and popular distributive functions, the two bureaus would likely find themselves in a powerful bargaining position, particularly in instances in which disputes went to the department's secretary.

The clear purpose in appealing to the secretary would be to reach a compromise acceptable to the FWS and its sister agency. The FWS might not like the need to compromise or the outcome of the compromise

itself, but there was little the FWS could do to avoid either. Awareness that the secretary would seek compromise further strengthened
the leverage of the two bureaus. If either persistently objected to a
proposed designation, the dispute would be given to a political appointee for resolution. For such political executives, the incentive is
to emphasize short-run achievements, symbolic actions, and cosmetic
panaceas.[42]

In the second instance, several proposed and final designations led
to lawsuits, heated controversy, and even to the agency's embarrass-
ment. As discussed previously, the most famous of the designations
involved the snail darter. Though the Tellico controversy received
considerable publicity, it was not the only cause of disenchantment
with the implementation of Section 7. Determination of critical habitats would be far more than the "simple, straightforward administrative action" that Keith Schreiner once claimed it would be.[43] Several
examples illustrate this.

The Grizzly Bear

Grizzly bears are among the largest and most ferocious mammals
found in the United States. Some exceed six hundred pounds. Their
size is not, however, the chief cause of concern. Many people believe
that grizzly bears cannot coexist with humans. Unfortunately for both,
they too frequently come into contact, often with death resulting for
one of the two species.

Grizzly bears once were found throughout the West, as far south
as Texas and as far east as Iowa and Minnesota. Today they are
found only in Alaska, Canada, and in or near several national parks
and forests in Montana, Wyoming, Idaho, and Washington. Some of
these parks, such as Yellowstone, are among the most popular and
frequently visited in the United States. These visits and the increasing demands placed on the parks create problems for the grizzlies.
Because of their size, each needs "large tracts of undisturbed range
simply to find the food to support its giant bulk." Research through
radio tracking has found that some grizzlies range over one thousand
square miles during their lifetime.[44]

In order to protect the remaining grizzlies outside of Alaska, the
FWS concluded that it was necessary to designate a critical habitat.
A further reason for designation involved the federal government's

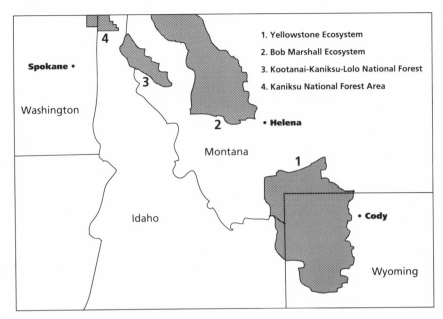

Figure 7.1   Proposed Critical Habitat for the Grizzly Bear. Source: *ESTB* 2
(Dec. 1976–Jan. 1977): 4.

ownership of most of the land the bears inhabit. As the FWS reasoned,
because most of the remaining 600 to 1,200 bears live on public
lands, early designation of critical habitat would be useful to the affected federal land managers in meeting their responsibilities under
Section 7. In response, the FWS proposed the designation of a critical habitat encompassing approximately thirteen million acres in four
states in November 1976 (see figure 7.1).[45] This would be the largest
critical habitat ever proposed—more than 20,000 square miles, about
half the size of Ohio. However large the proposed habitat might seem,
it represented less than 2 percent of the bear's original range. Some
people within the FWS (and many in the environmental community)
believed that the thirteen million acres would be inadequate. Some
groups wanted nearly twice as much area designated.

Rather than proposing only those areas known to contain bears, the
FWS also included adjacent areas that could sustain the grizzlies. In
each instance, however, rather than excluding areas where the bears
were not found, the FWS relied on administrative convenience. All the
proposed boundaries coincided with readily identifiable manmade or

geographic features. This did not make scientific sense, but it made a great deal of sense to bureaucrats concerned with trying to provide legal descriptions and easily defined boundaries.

In its proposal, the FWS stressed that the critical habitat would provide all the physical, biological, and behavioral requirements necessary for the species' survival, continued existence, and anticipated recovery. Although the proposal was based on what the FWS considered to be the best available biological data, the agency admitted that much additional research would be necessary. In other words, even without full knowledge of such things as the effects of land development or habitat-use patterns, the FWS was still willing to proceed.

This stance reflected an important shift in attitude. The FWS had previously emphasized that it would not determine any critical habitats until "every shred of biological data" had been obtained and analyzed.[46] The FWS did not have all the information it would have liked about grizzly bears, but designation of a thirteen-million-acre habitat was nonetheless deemed essential. Anything less than that size would require much more stringent regulation of public land use while the failure to designate any critical habitat was judged to be an unacceptable alternative. Until further information became available, the FWS favored a conservative approach that benefited the species. Such an approach, said the FWS, "will be mandatory for Federal agencies, regardless of the pressure generated by the special interest groups dedicated to maximum returns from their investments in developing public lands and public resources."[47] In taking this position, the FWS placed concern for the species above other, human concerns. This would be a risky strategy and a difficult one to sustain politically.

The wisdom of the strategy was quickly challenged. The bears' location on public lands affected the USFS, the NPS, and the BLM. All three agencies have close alliances with private users of these public lands, and the FWS clearly recognized the potential for controversy. The Service noted that most of the species' habitat is on public land, "which brings it into conflict with the vast and politically powerful economic interests which have traditionally developed and exploited public lands."[48] These interests include logging ventures, farmers and ranchers grazing livestock, developers of gas, oil, and mineral deposits, and those dependent on tourism and big game hunters. All these interests might have their activities restricted or regulated because of designation. Not surprisingly, these interests vehemently protested the proposed designation. Several public hearings attracted hundreds of people, most of whom strongly opposed designation.

All this pressure was supposedly irrelevant to the biological decision the FWS was expected to make. The pressure was not limited, however, to the public or affected economic interests. In rapid succession, the USFS, the BLM, and the directors of each of the four states' fish and game departments expressed their opposition. Several U.S. senators did the same. One of them arranged for the Senate Appropriations Committee to conduct a special public hearing just a few miles from part of the proposed habitat in Yellowstone National Park.[49] Such a hearing is rare, since the Appropriations Committee does not have any oversight responsibility for the FWS or the Endangered Species Act. Nonetheless, the committee did have considerable control over the agency's budget, and the hearing signaled intense congressional displeasure with the FWS. When the FWS needed institutional allies, there were none to be had. The agency opponents to the grizzly habitat were responding to their constituents, and they found themselves united in opposition to the FWS.

Not only did the FWS have to handle this opposition, but it also had to address several problems of its own creation. One contentious issue involved the restrictions that might be imposed on use of the proposed critical habitat. The FWS attempted many times to assure people that designation would affect only federal agencies and their activities. These assurances were not always successful. One local newspaper blasted the proposed designation because it would figuratively place a barbed-wire fence around the critical habitat and then lead to economic disaster as tourism ended.[50] Senator James A. McClure (R.–Idaho) charged that the habitat "would become a permanent headlock on governmental agencies in the performance of their duties."

The FWS found itself frustrated in attempting to address these complaints. At one of several public meetings, for example, the FWS assured the public that a critical habitat for the grizzly bears was unlikely to disrupt their lives. A few minutes later an FWS spokesman said it would be up to federal land managers to decide what could or could not be done and which activities would be curtailed or allowed to expand within the habitat.[51] Undermining its argument that only biological data would influence its decisions, the Service argued that public opinion was on its side because over five hundred people had written letters supporting the previous year's listing of the grizzly bear.

After making the case that determination of critical habitat for the grizzly bear was essential, the FWS waited three years before acting. The proposed designation was withdrawn in 1979. No subsequent efforts to establish a critical habitat for the bears have since been initiated. At the least, the FWS suffered a blow to its administrative credibility when it withdrew the proposal. Public opinion rather than scientific data ultimately played the key role in the outcome. Hundreds of people supported the bear's listing, but they were spread throughout the United States. They had no economic incentive and limited ability to mount an effective campaign for the critical habitat. Just the opposite situation occurred for opponents. They were geographically concentrated, and they believed that a critical habitat would imperil their economic well-being. This situation is not at all unusual in regard to the protection of endangered species—diffused support, but focused opposition to their protection.

The FWS also encountered problems with its proposal because it could not specify the consequences of designation. This caused considerable anxiety and served to underscore the need for regulations governing the use of Section 7. Further evidence of all these problems became available when the FWS proposed the establishment of a critical habitat for the Houston toad.

The Houston Toad

The Houston toad is rarely sighted because it lies torpid in sandy soil during much of the year, usually from June through January. The toads appear above ground during their brief breeding seasons in the spring of each year, but can be located only at night during periods of high humidity or persistent rainfall. Even in the best of these conditions, the toads are still difficult to locate and are among the rarest and most critically endangered amphibians in the United States.

As a result of biological surveys, the species was listed as endangered in 1970. A few years later a consultant to the FWS recommended that critical habitat for the species be designated in parts of three counties in southeast Texas. One of these counties, Harris, includes the city of Houston. After reviewing the consultant's report and the proposed boundaries of the habitat, officials at FWS headquarters asked its regional office to redefine the proposed boundaries because they posed "serious difficulties."[52] The FWS justified this request on the basis of what it assumed to be errors in the consultant's data.

The regional office sided with the consultant's recommendations and reported its surprise with the reaction from headquarters. The tone of the headquarters' statement, said the acting regional director, "is that politics may make designating this area as critical habitat difficult. Although this is undoubtedly true, now does not seem to be the time to raise the problem."[53] The Service's headquarters retreated, and it proposed the designation of a critical habitat for the toads in May 1977. In proposing the two rural sites as well as seven sites in metropolitan Houston, the FWS called attention to its belief that much "of the hope for survival and recovery for this species depends upon maintenance of suitable, undisturbed habitat and breeding sites."[54] Furthermore, the final designation of habitat would be based, the Service righteously proclaimed, solely on biological factors as determined by the best available scientific data. To do otherwise would violate the law.

However strong the desire to rely solely on scientific data, the FWS would also have to contend with public opinion and a metropolitan area obsessed with growth and alteration of habitat. At the time of the proposal, Houston was the fastest-growing major city in the United States, with about one thousand people moving there each week. As an indication of its commitment to unrestrained growth, the city had no planning board and no restrictions on land use. Landowners could develop their property any way they wished.

With its rapid growth, much of Houston's landscape was also changing. At one time all seven of the proposed sites for critical habitat in Harris County probably had served as the toad's habitat, but this was no longer true in 1977. Indeed, the FWS inadvertently invited ridicule when it was pointed out that one of the sites in Harris County contained a shopping center, an industrial park, and several large apartment complexes. No toads had been sighted there, and existing developments made much of the site completely unsuitable for the toads. Some developmental activities occurred on several of the other six sites. Only one toad had been seen on one of the sites in the previous ten years.

The Service's efforts to designate critical habitat within Harris County received extensive coverage in local newspapers and even national media attention. Due to the criticism directed at its proposal, the FWS convened a team of biologists to review each of the proposed sites. The review team had no qualms about recommending the two rural sites. Of the seven sites in Harris County, the team recommended to the FWS regional director that two be dropped from further consideration. The team's toad experts agreed unanimously, however, that four areas in Harris County should be designated as critical habitat. The experts all agreed that the areas were also necessary for the species' survival and recovery.[55] The experts were less certain about a fifth site, but they still believed it to be critical. The team further recommended that six other areas in Harris County be considered as additional critical habitat.

In his recommendation to FWS headquarters, however, the acting regional director in Albuquerque hedged. He claimed that if the five original sites in Harris County were designated as critical habitat, the public outcry would be immediate and vociferous because the FWS would be unable to show that toads had used any of the sites within the previous ten years. "We understand the final critical habitat determination must be based upon biological data rather than political or economic concerns," he added, but we "feel our biological data is presently not sound enough to effectively counter these pressures."[56] This was a telling admission since it not only acknowledged the relevance of nonbiological factors, but it also cast the FWS as obligated to be sensitive to public opinion. Supposedly, if the biological data justifies the designation of a critical habitat, then public opinion is an irrelevant consideration.

In fact, however, public opinion became an important consideration, because the FWS was unsuccessful in conveying the consequences of designation. As noted earlier, Section 7 prohibits only federal agencies (and activities they fund) from adversely affecting a species or its critical habitat. Many people in Houston were convinced, however, that designation would create absolute barriers to all private development.

With this public pressure and the recommendation from its regional director, what the FWS did next was not unexpected. The two rural sites but none of the seven sites in Harris County were designated. As the Service explained, the decision not to designate the sites in Harris County "was based largely on the findings of a special review team organized by the Service. . . ."[57] The FWS said that further surveys would be conducted at the undeveloped portions of five of the most promising sites in Harris County, again suggesting that biological data would ultimately determine FWS action. In an internal memorandum, however, the associate director, Keith Schreiner, emphasized that the surveys were to be more than just biological assessments of the toad's habitat. He said that a research plan "should be so devised as to gain the full support of business persons in the proposed areas. There will be *no* benefit to the species until compromise is made by *all* parties."[58] What was not clear was how the FWS expected scientific data to be compromised.

A rather startling directive followed the compromise directive. The FWS instructed its regional office in Albuquerque to acquire Ellington Air Force Base, which was part of one of the five sites to be studied in Harris County. A bewildered response, perhaps with tongue in cheek, quickly came from the regional office: "We are unsure as to how to go about acquiring Ellington Air Force Base, but will maintain contacts with people in Houston. If and when the AFB comes up as surplus property we will attempt to acquire it."[59]

The regional office complied with the request for additional surveys, but these were inconclusive. No areas within Harris County have been designated as critical habitat. As one observer of the situation later stated, it is difficult to determine whether politics or science was preeminent in the Service's decisions about the species' habitat.[60] No doubt science was involved, but agency officials were aware of and sensitive to community opinion. This opinion seemed to weigh more heavily than did the conclusions of the hand-picked team of biologists.

The two case studies have many similarities, but they were not limited to the bears or toads. Controversy and legal action surrounded the designation of a critical habitat for the Mississippi sandhill crane. In its designation of a critical habitat for the California condor, which requires almost total wilderness to reproduce, the FWS was accused of establishing boundaries that would accommodate proposed phosphate mining on land the BLM administers. For two butterflies in California, the boundaries for their proposed habitats were allegedly drawn so that they would not interfere with operations at the Los Angeles International Airport, an oil refinery, and a rifle range.[61] In short, there was controversy associated with many of the initial efforts to determine critical habitats. These controversies focused as much on the proper boundaries of the habitats as they did on the consequences of designation. A flurry of court decisions—involving the snail darter, the whooping crane, the sandhill crane, and the Indiana bat—heightened federal agencies' concern about the consequences of further designations. Each of these decisions, but most important, the

Supreme Court's verdict in *TVA* v. *Hill,* reemphasized federal agencies' obligations under Section 7.

There was increasing concern that future designations would lead to further conflicts, and the FWS did little to allay this apprehension. Based on FWS projections, the number of additional proposals and final designations would mushroom in future years. The FWS was encouraged to designate still more critical habitats when President Carter directed that a survey of all federal agencies be conducted in order to identify "within the shortest possible time" areas that should be designated.[62] The president also directed Interior to expedite the designation of additional critical habitats in order to protect listed species. The FWS responded enthusiastically to the president's message. In 1978, as an illustration, the FWS said that it anticipated the designation of 293 additional critical habitats before 1980.[63]

The FWS seemed poised for further action, but it still had not satisfied other federal agencies that wanted more specific direction about how they were to comply with Section 7. Indeed, several judicial decisions had interpreted the section in a manner much different from FWS implementation. Thus, providing further guidance to federal agencies was the task to which the FWS next turned its attention.

One of the complaints directed at the FWS concerned its delay in providing regulations for the implementation of Section 7. The Service had issued guidelines in early 1976, but these provided recommended procedures rather than binding rules. Considerable uncertainty thus existed about the consultation process. After two years' experience with the guidelines, an internal appraisal revealed how ineffective they had been: "The limited consultation that took place often occurred in an ad-hoc and haphazard fashion."[1] This was surely not a positive evaluation of guidelines intended as a starting point for the development of regulations, but far more important, for the protection of endangered species.

Rather than taking the lead in proposing regulations, the FWS deferred to the OMB, which intervened. The OMB asked to review the guidelines, ostensibly to provide other agencies a chance to participate in the writing of the final rules. After two such reviews and after assessing reactions from other agencies, the FWS proposed regulations for the implementation of Section 7 in early 1977.[2] Despite the explicit statutory requirement for consultation, the proposed regulations indicated that consultation would still be discretionary. A project agency would decide whether it wanted to consult with the FWS, an approach that several agencies strongly favored. These agencies clearly wanted to retain authority over their projects and could not reasonably be expected to be too enthusiastic about FWS participation or the uncertain outcome associated with the consultation process.

Opponents of the proposed regulations cited not only the mandatory nature of the consultation process, but also two relevant court cases: *National Wildlife Federation* v. *Coleman* and *TVA* v. *Hill*.[3] In both cases, U.S. courts of appeals had affirmed that consultation is mandatory before any federal agency can take an action that might adversely affect listed species or their critical habitats. As a result, when the final regulations on Section 7 were published in January 1978 (more than four years after the passage of the Endangered Species Act of 1973), the FWS changed its long-standing position. Noting that the courts' decisions were binding, the FWS said that consultation would henceforth be mandatory.[4] What did not change was the action agency's authority to decide whether to proceed with a project *after* consultation had occurred. Even in the face of a finding that such an action would jeopardize a species' continued existence or its habitat, the action agency would retain the right to make the ultimate decision about a project's fate.

The proposed regulations had stipulated, in a conciliatory manner, that the degree of a project's completion could also determine whether consultation should be initiated. It was not the intent, the FWS stated, to use Section 7 to "bring about the waste that can occur if an advanced project is halted." Furthermore, the affected agency would be able to "decide whether the degree of completion and extent of public funding of particular projects justify an action that may be otherwise inconsistent with [S]ection 7."[5] Such a statement obviously represented the Service's deferral to the judgment of other agencies. More important, the proposal revealed the apparent willingness of the FWS to compromise a species' continued existence in instances in which an action agency deemed that it had spent more than it would like to "lose," however small that amount might be. The FWS again reversed itself in the final regulations because of the Supreme Court's decision in *TVA* v. *Hill*. In this case, which involved the snail darter, the Court ruled that a project's degree of completion was irrelevant, and that extinction was not a variable that could be added to a calculation of a project's costs and benefits. No matter how much money had been spent on a project, the amount provided insufficient justification to forego consultation. This reasoning pleased those in favor of preservation, but it meant that a newly discovered or recently listed endangered species might halt or delay the completion of any federal project.

In still another area, the FWS reversed itself because of the TVA decision. The final regulations specified that, until the consultation process was completed, a federal agency would be precluded "from making an irreversible or irretrievable commitment of resources" that would foreclose the remedial or alternative measures being discussed.

In response to suggestions from two federal agencies that nonbiological variables be included in decisions to designate critical habitats, the FWS responded quite negatively. After having included such variables in several of its earlier actions, the FWS self-righteously asserted that in the designation of critical habitats, the

> focus is entirely on the biological and ecological needs of the listed species. The consideration of various socioeconomic factors is irrelevant in determining what these biological needs are. The inclusion of socioeconomic considerations would diminish the effectiveness of conservation programs for the recovery of listed species by distorting the estimate of its true habitat needs.[6]

The FWS similarly rejected suggestions that it provide more explicit definitions of jeopardy, critical habitat, and adverse modification or destruction of critical habitat.[7] The FWS maintained that flexibility was necessary in order to account for the unique needs of many endangered species. The FWS also reemphasized that critical habitats could include the present range of a species as well as "additional areas for reasonable population expansion."

These regulations softened the growing demands for clarity, but the regulations did not quiet those critical of the implementation of Section 7. These critics included those who complained that the FWS was too timid in its actions as well as those who asserted it was being far more aggressive than the law allowed.

The FWS is Too Timid

Several environmental groups faulted the FWS for inadequate implementation of Section 7. These critics claimed that far too few critical habitats had been designated, that many of them were too small, and that the Service was not meeting its statutory obligation regarding consultation. In reauthorizing appropriations for implementation of the Endangered Species Act in 1976, as an illustration, a congressional committee complained that not a single critical habitat had been designated more than two years after passage of the act. The delinquency, said the committee, was due to shortages of staff and budget.[8] Eight OES staff members were not only responsible for listing species but were also expected to handle the designation of critical habitats.

Lack of sufficient financial resources affected their ability to do so, so the committee recommended a "modest increase" in funding for the FWS.

Environmental groups were equally perturbed at the FWS, claiming that it was being far too cautious in its designation of critical habitats. The Defenders of Wildlife objected to designations of "postage stamp size necessary to continue the survival of the last handful of any species."[9] To remedy this deficiency, Defenders advocated that habitat necessary for survival and recovery should be considered critical "until the agencies which are fostering and promoting development can prove that such habitat is not critical and necessary for the survival, welfare, and subsequent restoration of the species." This approach would shift the burden of proof from the FWS to the action agency.

Other complaints addressed Interior's failure to comply with the law's requirements and frequent administrative delays in the consultation process. One part of Section 7 required the department's secretary to ensure that the programs he administered were used to achieve the purposes of the Endangered Species Act. The secretary thus instructed all units in the department to review their programs to see how they might be affecting listed species, but not all Interior agencies did so. The BOR failed to disclose that some of its projects in California had reduced the remaining habitat of at least three endangered species.[10] Some Interior units, including the FWS itself, identified potential conflicts with endangered species but had not acted promptly to resolve them.

Section 7 was not intended to hinder federal projects. Nonetheless, an agency conscientiously attempting to comply with the section could quickly discover that compliance might delay a project. When the FWS began the consultation process, it said that biological opinions would be provided to agencies with which it consulted. These opinions would outline how a project might affect listed species or their habitats. Federal agencies would reasonably expect to receive these opinions soon after consultation had been initiated; long delays would discourage further cooperation. However desirable prompt responses might be, the FWS was frequently unable to provide biological opinions in an efficient manner. In many instances the opinions were not available within the sixty-day time limit that the FWS had informally established for itself. Some opinions were delayed for several months either because of the complexity of the consultation or because of the shortage of FWS staff and resources.[11] At one time the Corps of Engineers requested a consultation to consider the possible effects of a proposed dam in Wyoming on whooping cranes. After waiting two months to respond to the request, the FWS said that the dam would affect the crane's as-yet-undesignated critical habitat. Before a final biological opinion could be issued, however, the FWS claimed that about three more years' research would be necessary. The Corps deemed the projected delay to be unacceptable. It went ahead and issued permits for the project, but successful legal action delayed further construction.[12] A compromise eventually allowed the dam's completion and protection of the crane's habitat. The length of delay in this case was atypical, but inordinate delays were the rule rather than the exception.

The prospects for delay undoubtedly discouraged some agencies from initiating consultations with the FWS. Moreover, Interior's own lack of compliance with Section 7 set a poor example for other agencies. Several units within Interior, including the FWS, did not initiate consultations in instances in which they probably should have. As the GAO later emphasized, if Interior, the agency with lead responsibility for the law, "justifies noncompliance with the act because of funding and staff constraints, other Federal agencies whose primary responsibilities are not necessarily compatible with the conservation of endangered and threatened species cannot be expected to adhere to the [S]ection 7 requirements."[13]

Critics were further concerned because the FWS had neither kept accurate records of the number of consultations that had occurred nor attempted to determine what had happened to endangered species after consultation. Not until September 1977 did the Service begin to keep records of consultations. The best that the FWS could do was to estimate that approximately 4,500 consultations had occurred during the period in which consultation had been discretionary.[14] This was a questionable estimate, based more on speculation than on conscientious record keeping.

Whatever the number of consultations, they reflected measures of administrative effort and were meaningless without some parallel consideration of the consequences of the consultations. The FWS could not accurately assess how many times agencies justifiably should have consulted but did not, how many times projects potentially jeopardized listed species or their habitats or, most notably, how agencies responded to recommendations from the FWS to mitigate or eliminate threats to endangered species. The ultimate purpose of Section 7 is not for agencies to consult, but rather to conserve and protect listed species and their habitats. For all but a few cases, primarily those that had been subject to litigation, only limited data were available about what had been done to protect endangered species subsequent to a consultation. Likewise, several agencies had to be reminded of their duties to consult while others ignored requests from the FWS to consult. Rather than pursue the issue, the FWS occasionally waited for a private citizen or environmental group to file suit to enforce the consultation requirement.[15] In short, the overall effectiveness of the consultation and no-jeopardy requirements were almost completely unknown. It was against this background (and the Service's frequent inability to identify the consequences of designating critical habitats) that Congress and many interest groups expressed their disillusionment with the implementation of Section 7.

The FWS is Too Aggressive

Just as the 1978 congressional hearings on reauthorization of the Endangered Species Act provided a chance to complain about the listing of species, the hearings also created an opportunity to challenge the implementation of Section 7. Several members of Congress were furious due to the designation of critical habitats, citing the snail darter, the absurdity of proposing a shopping center for the Houston toad's critical habitat, and so on. As one representative generalized, the OES "has gone too far in just designating territory as far as the eyes can see and the mind can conceive."[16]

When members of Congress considered the number of species likely to be listed, the number of anticipated habitat designations, and the number of consultations the FWS projected for the coming fiscal year (approximately 20,000), the result was considerable alarm. Nonetheless, most members of Congress expected most consultations to proceed without controversy and without significant effect on federal activities or endangered species. If a project endangered a particular species, the species could be transplanted to a new location, or so Congress seemed to believe. Using endangered plants found in isolated areas in Hawaii and California as examples, the House Committee on Merchant Marine and Fisheries naively declared that if a conflict with a federal activity should develop, it would "be a simple matter to relocate populations of these species to avoid a violation of

[S]ection 7."[17] Obviously, the basis for such an assumption was not necessarily a realistic one. What the committee did not seem to appreciate is that certain species are found only in isolated areas because other locations do not provide suitable habitats.

Even if most consultations did not produce controversy or could be resolved without detriment to an endangered species, concern still existed that application of Section 7 would halt or delay many federal projects. Advocates of changes in the Endangered Species Act certainly wanted to leave this impression, and some members of the Senate Committee on Environment and Public Works accepted this assertion as if it were a fact. In its consideration of changes to the 1973 law, the committee noted "that a substantial number of Federal actions currently underway appear to have all the elements of an irresolvable conflict within the provisions of the act."[18] This number, added the committee, might increase significantly in the future, as more species were listed and critical habitats designated.

To remedy the perceived deficiencies in Section 7, both the Senate and the House considered many amendments in 1978. Some of the proposed amendments were intended to allow an exemption from the consultation and no-jeopardy requirements for all projects under contract "or otherwise underway," or for those more than 50 percent completed. Still other proposals would: a) allow a sponsoring agency to balance a project's social and economic benefits against the possible "esthetic, ecological, educational, historical, recreational, or scientific loss to the public" that would occur if a species were to become extinct; or, b) require the FWS to consider the social, economic, scientific, ecological, archeological, and national security impacts before designating any critical habitat. If either proposal had been approved, the OES would have to expand considerably its staff or designate critical habitats in only the most unusual circumstances. Congressional and presidential unwillingness to provide adequate funding suggested that the latter alternative would be a far more likely result.

After considering many different versions of a new Section 7, which governed interagency cooperation, Congress made many important changes in existing procedures. The original Section 7 in the 1973 law included only four sentences, but the revised section included more than six, single-spaced pages of text. The Endangered Species Act Amendments of 1978 retained and largely restated the existing consultation and no-jeopardy rules that the FWS had issued in early 1978. In regard to critical habitats, however, the new law also:

1. Defined critical habitats and specified that their designation be based on the best scientific data available;
2. Indicated that critical habitats are *not* to include the entire geographical area that an endangered or threatened species could occupy, unless a determination is made that such areas are essential to a species' conservation;
3. Required specification of critical habitats "to the maximum extent prudent" at the same time a species was proposed for listing;
4. Required consideration of economic and other relevant impacts before designation of a critical habitat;
5. Required a "brief description and evaluation" of activities that might modify a designated critical habitat or that might be affected by designation;
6. Required that procedures similar to those used to list a species also be used to designate critical habitats including formal notification in the *Federal Register* and opportunities for public comment and public meetings;[19] and
7. Allowed the exclusion of areas from a critical habitat when the benefits of exclusion outweighed the benefits of including the area.

In regard to consultation, Congress added several provisions to clarify the process and the obligations of each agency involved in it. The new provisions required a federal agency to prepare a biological assessment indicating whether its project would affect a listed species or one that had been proposed for listing. Consultation would be required if this assessment revealed that the project might jeopardize a listed species' continued existence or adversely affect its habitat.

Once consultation was initiated, the revised law stipulated that the process had to be concluded within ninety days, unless an extension was mutually acceptable to the FWS and the action agency. During the consultation period, the action agency would be precluded from making any "irreversible or irretrievable commitment of resources" that would foreclose the implementation of any "reasonable and prudent measures" designed to mitigate any adverse effects.[20] This requirement was almost identical to the restriction on the commitment of resources found in existing FWS regulations on Section 7. Last, the amended law obligated the FWS to provide a written biological opinion at the conclusion of the consultation. This opinion would indicate whether a project would violate Section 7 and, if so, what reasonable and prudent alternatives the action agency might consider.

By changing the definition of what constitutes a critical habitat and by modifying slightly the procedural requirements associated with consultation, Congress guaranteed that the FWS would once again have to prepare new regulations for Section 7. Congress did not facilitate this process; more changes were made to the law in 1979. One of these changes modified the no-jeopardy requirement; in its place, Congress required agencies to ensure that their activities are "not likely to jeopardize" a listed species or its critical habitat.[21] Additional changes were made in 1982, when Congress again amended the Endangered Species Act. The latter amendments allow: a) the participation of applicants for federal permits or licenses in the consultation process; and, b) a federal agency or a permit or license applicant to engage in the incidental taking of an endangered species, provided that the taking does not jeopardize the species' continued existence.

The revised statutory provisions may have attempted to reduce delay, but such intentions did not ease efforts to produce new regulations explaining to federal agencies what their consultation obligations would be according to the 1978, 1979, and 1982 amendments. In fact, proposed regulations governing Section 7 were not issued until mid-1983, and the final rules were not published until June 1986, nearly eight years after the major changes introduced in 1978.[22]

One of the more controversial features of the new regulations involved the definition of jeopardy. The initial definition, which had been included in the regulations published in January 1978, had been based on a desire for flexibility and the need "to deal with every possible . . . situation" in such a way as to benefit listed species.[23] Although the definition of jeopardy provided in the 1986 regulations was almost identical to that in the earlier rules, flexibility seemed much less important in 1986. The new regulations emphasized that a project would have to have a detrimental impact on both the survival *and* recovery of a listed species in order for the FWS to conclude that the project jeopardized the continued existence of a species.[24] Some people had wanted the FWS to issue jeopardy opinions whenever any federal action impairs a listed species' survival *or* recovery. The FWS responded unfavorably to these suggestions, arguing that its choice did not misconstrue the act's purposes, as some people believed to be the case. These people cited the requirements of the Endangered Species Act that federal agencies conserve endangered and threatened species and use whatever methods are necessary to recover such species. Accordingly, the advocates of change said it would make no sense to allow such agencies to undertake projects that might vitiate recovery efforts. Despite this reasoning, the FWS claimed it did not have the legal authority to compel other federal agencies to conserve listed species or to issue jeopardy opinions when a proposed action would adversely affect only a listed species' recovery, but not necessarily its survival. This was a fine distinction, the FWS admitted, because impairment of recovery efforts could risk a species' continued existence.

With the issuance of the 1986 regulations governing the implementation of Section 7, the FWS had responded to calls for greater clarity and specificity. Not all federal agencies were satisfied with the regulations, but at least they formally and finally knew what was expected of them in regard to the consultation process. From the perspective of the FWS, the regulations represented the codification of standard operating procedures, which are vital to any administrative agency's success.[25] These regulations achieved the Service's organizational goals, but the regulations largely addressed procedural rather than substantive concerns. The regulations indicated how consultation should occur and the procedures that the Service would employ in deciding whether to issue jeopardy opinions. Such guidance is essential for efficient administration of the Endangered Species Act. The important issue is, however, not what procedural accomplishments there have been, but rather the consequences of designating critical habitats and of Section 7's consultation requirements. Are the habitats of listed species effectively protected as a result of the Endangered Species Act? Does consultation improve federal agencies' responsiveness to the needs of endangered and threatened species? How assertive is the FWS in its efforts to protect species and their habitats? How responsive are federal agencies to recommendations about how their projects can or should be altered to minimize or eliminate threats to listed species? Answers to these questions are crucial to any determination of the effectiveness of the Endangered Species Act. However important the questions are, regulations on consultation do not address any of them. In part, this occurs because the FWS does not have the authority to require federal agencies to report what actions they have taken in situations in which the Service has issued a jeopardy opinion or has provided reasonable and prudent alternatives to a proposed action. Consequently, other data must be examined in order to assess efforts to protect endangered and threatened species.

New Perspectives, New Approaches

Congressional scrutiny of the endangered species program was especially intense in the late 1970s and early 1980s. From November 1978 through October 1982, Congress approved at least five laws affecting the protection of endangered species. If there is a consistent theme in these efforts, it is surely that no federal agency, including the FWS, should have a free rein to manage or protect endangered species. Congress wanted to be sure that programs to protect endangered species are responsive to the concerns of its members. Similarly, members of Congress want to ensure that favored projects, like those within their congressional districts, do not suffer from overly aggressive implementation of the Endangered Species Act. The 1973 act allowed the FWS to list species and designate critical habitats with only a minimal level of congressional guidance. That law, one might argue, reflected trust in the FWS's objectivity and scientific competence. Although subsequent laws have not completely repudiated this trust, Congress made clear that both procedural and substantive changes were necessary in the listing of species, the determination of critical habitats, and in the consultation process. The changes also made certain that the protection of endangered species would no longer be the preeminent value in policymaking.

For the FWS, the messages inherent in the legislative flurry were clear and distinct. The 1973 act encouraged the agency to be an aggressive advocate for listed species. Subsequent legislation and congressional unwillingness to provide adequate resources undermined this approach, and a review of the FWS's actions provides evidence to support this conclusion.

The Determination of Critical Habitats

In the mid-1970s, after the FWS defined a critical habitat, the Service announced its intention to designate habitats for *all* listed species. The FWS declared it essential that each listed species have its habitat specified in order to provide the full protection of the law. By early 1978, critical habitats had been designated for twenty-six species, proposals for thirty-eight species had been issued, and proposals were being prepared for another eight. This meant, however, that more than 130 additional habitats would have to be designated in order to accommodate other native species already listed.

Table 8.1   FWS Designations of Critical Habitats for Priority Species

Priority Category in 1978	Number of Critical Habitats		
	To be Designated	Proposed	Actually Designated[a]
Extremely High Threat	8	0	0
High Threat	40	3	3
Medium Threat	46	2	1
Low Threat	41	0	0
Totals	135	5	4

Source for priorities: FWS and NMFS, "Critical Habitat Priorities," 17 Feb. 1978.
Note: The table excludes species that have been delisted since 1978 as well as those that are the sole responsibility of the NMFS.
[a]Through 31 May 1990.

Rather than designate habitats for these species in a random order, the FWS prepared a priority list of intended designations in early 1978.[26] The list, which included four priority categories, was based on the degree of threat believed to confront each species. For the species for which the FWS had responsibility, thirty-eight faced threats that substantially jeopardized their continued existence. These species were given the highest priority, and prompt designation of their critical habitats was considered imperative: "Such species may be undergoing a precipitous, continuing population decline, or face imminent threats which will essentially destroy all or a major part of their habitat. Extinction is likely in the immediate future unless measures are taken to develop a recovery program, including determination of Critical Habitat."[27]

For another fifty species facing a medium threat, a temporary delay in designating critical habitats would be tolerable. For forty-three species facing low threats, designation could be delayed for several years. Finally, nine species were said to be so rare and to face such an extremely high threat that determination of their critical habitats was deemed to be infeasible. Should remnant populations of any of these nine species be found, the FWS said that it would use its emergency authority to designate critical habitats as quickly as possible. Altogether, the FWS planned to designate more than 110 critical habitats between October 1977 and October 1979.

As the data in table 8.1 reveal, efforts to use the 1978 priority system for critical habitats have been meaningless. Of the 135 species on that priority list that were still listed in 1990, the FWS has designated habitats for only 4 species.[28]

Several possible explanations are available for the dearth of designations. As already noted, the 1978 amendments required assessment of the economic and other consequences of designating a critical habitat. This requirement affected scores of proposed designations and caused the FWS to withdraw pending proposals for nearly seventy habitats.[29] No doubt, the limited OES staff and the absence of appropriate expertise in economic analysis precluded the preparation of many other proposals. Strong opposition to many of the Service's initial designations also discouraged some activity.

More recently, there has been a lack of consensus within the FWS about the utility of and need for critical habitats. The legislative history of the 1978 Endangered Species Act Amendments reveals that Congress expected critical habitats to be designated in all except rare circumstances, such as when designation would be disadvantageous to a species.[30] The result is that for many species, especially plants, the FWS makes no effort to delineate a critical habitat. Designation of a habitat requires the publication of detailed maps showing the exact location of a threatened or endangered species. For plants, such maps could encourage vandalism or collection of a species so only occasionally are critical habitats designated for plants.

Mobile species are less prone to capture than are plants, but this has not increased the number of designations for the latter. Indeed, considerable disagreement exists among FWS scientists about the advantages of determining critical habitats for all species. Opponents of designation argue that failure to determine a critical habitat in no way diminishes the protection available to listed species. The absence of a formally designated critical habitat does not prevent the issuance of jeopardy opinions for projects that might destroy or adversely modify a species' habitat.[31]

From a practical perspective, designation of critical habitats is among the most controversial FWS activities. On this point one FWS biologist noted that "nothing buys us as much opposition as critical habitats. They are more trouble than they are worth."[32] When asked why any critical habitats are proposed, the biologist provided a straightforward response: "It beats the hell out of me!"

Critical habitats are intended to assist federal land managers in identifying the locations of listed species. With this knowledge managers can determine when consultation with the FWS is required. This was an initial expectation, but the public perception, notes the FWS, is that critical habitats "curtail or forbid all human activities within the area so designated."[33] This has never been the case, but some critical habitats encompass private property. Some landowners thus complain that designation represents an unconstitutional "taking" of private property without compensation. This complaint is without legal merit, but here perception is more important than reality.

According to the FWS biologist, critical habitats have a poor public image, do not benefit listed species, and often create unnecessary risks for many species. In one instance, as an illustration, the FWS proposed the designation of a critical habitat for the Chihuahua chub, a species of fish. The proposal was eventually withdrawn because the Service received so many threats from people claiming they would deliberately destroy the chub if its habitat was formally designated.[34]

Appropriate boundaries are also difficult to determine, and in the biologist's view, the size of some critical habitats frequently depends less on scientific data than on personal preferences and administrative convenience. When the FWS proposed a habitat for the large-flowered fiddleneck, a rare flower found only in California, the area exceeded the plant's then-current range in order to accommodate expansion or relocation.[35] In contrast, when the FWS designated a critical habitat for the Coachella Valley fringe-toed lizard in southern California, the Service was exceedingly circumspect. After noting that approximately 100 square miles of suitable habitat remained for the lizards, the FWS designated a habitat of slightly less than 19 square miles.[36] In the first case, all the flowers are located on land the Department of Energy owns, and the department did not oppose designation. Protection of the lizards has been controversial because much of their habitat is privately owned and is subject to intense developmental pressures. Some developers claimed that designation of a "large" critical habitat for the lizards would cause a loss in revenues of nearly $25 million per year.[37]

All these problems have caused many people in the FWS to oppose further designations, but this view is not universally shared. The consequence of this divided opinion is that some species are more likely than others to have critical habitats designated, solely because of differences in opinion among FWS staff. From the beginning of 1979 through 1989, as an example, seventeen species of birds were listed but only one critical habitat was designated for them. Over the same time period, forty-five fish species were listed. Of these, thirty-one

Table 8.2   Designation of Critical Habitats by Class

Taxa	No. of Native Species Listed	No. with Critical Habitats	% with Critical Habitats	% of Habitats Designated	
				Before 1979	After 1979
Mammals	57	12	19.3	33.3	66.7
Birds	82	10	12.2	90.0	10.0
Reptiles	30	13	43.3	53.8	46.2
Amphibians	11	3	27.3	66.6	33.3
Fishes	83	38	45.8	18.4	81.6
Crustaceans	9	1	11.1	0.0	100.0
Insects	18	5	27.8	0.0	100.0
Arachnids	3	0	0.0	NA	NA
Snails and Clams	43	0	0.0	NA	NA
Plants	218	25	11.5	8.0	92.0
Totals	554	107	19.3	28.0	72.0

Note: The data include all native species listed as either endangered or threatened on 31 December 1989. The critical habitat designation for the snail darter was rescinded in 1984.

had critical habitats designated. As the data in table 8.2 confirm, considerable variation exists among different classes. Similarly, the data indicate that birds, reptiles, and amphibians had most of their critical habitats designated before passage of the 1978 amendments. Fishes and plants have received more attention since 1978.

It is also the case that the time involved in the designation of critical habitats increased significantly after the passage of the 1978 amendments. The amendments require an analysis of economic and other impacts before a critical habitat can be proposed for designation. This process has increased the time necessary to *develop* a designation proposal. This requirement does not explain, however, why it has taken the FWS 70 percent more time to move from a proposed designation of a critical habitat to a final rule since the passage of the 1978 amendments. Since the amendments' passage, the average length of time from proposal to final designation has increased to more than seventeen months from slightly over ten months (see appendix table B.1). The differences in the average designation times are especially pronounced for fishes, mammals, and reptiles.

In sum, critical habitats have had a checkered history. The FWS initially announced its intentions to designate critical habitats for all listed species and to do so as quickly as possible. These expectations just as quickly evaporated in the face of budgetary restrictions, public and congressional dissatisfaction with the consequences of designation, and perhaps most important, a revised perception about the relative desirability of the entire effort.

These variables similarly appear to have affected the FWS's planning efforts, as reflected in its stated priorities. In one of its early announcements, in 1975, the Service identified ten high-priority species for which it intended to determine critical habitats as "rapidly as possible." [38] Nearly fifteen years later, designations had not even been proposed for three of the ten. In a comprehensive listing of priorities issued in 1978, the FWS listed fifteen species of birds and twelve species of clams that might face immediate extinction unless their habitats were designated quickly. No critical habitats have been proposed for these twenty-seven species. The number of habitats designated has increased, at a rate of slightly less than eight per year since 1976, but the habitats designated reflect neither a series of well-ordered priorities nor a comprehensive plan for protecting species' habitats, at least through the designation of critical habitats.

The Acquisition of Habitats

In addition to designating critical habitats, the FWS can also attempt to protect habitats through its access to money from the LWCF. Although the 1966 and 1969 endangered species laws limited the amount of money that could be spent to acquire habitats, the 1973 Endangered Species Act removed the limitation. The FWS could henceforth receive as much money to acquire habitats of listed species as Congress was willing to provide.

This new arrangement does not mean, however, that the FWS gets as much money as it might prefer. In order to receive funds to lease or purchase habitats (or to acquire easements to them), advocates of endangered species must compete not only with other claimants within the FWS, such as the National Wildlife Refuge System, but also with other, more powerful federal agencies, such as the USFS, the BLM, and the NPS. For the latter agency, the LWCF is its only source of money to acquire additional parklands. This may explain why the NPS, an agency that "distributes" benefits, received almost 70 percent of all the money Congress appropriated to federal agencies from the LWCF between fiscal years 1976 and 1985. Put in other terms, the Park Service's share of the funds is larger than the combined shares of the USFS (approximately 20 percent of total appropriations), the FWS (approximately 11 percent), and the BLM (approximately 1 percent).[39]

Within the FWS, the endangered species program has had a mixed record of success in gaining support from the LWCF. In some years the FWS has requested that more than half of its congressional appropriations be used to benefit listed species (see table 8.3). In other years, endangered species have suffered because of proposed budget reductions, as in 1981. For that fiscal year, Congress provided more than 90 percent of the Carter administration's request for the acquisition of habitat of listed species. After taking office, however, the Reagan administration rescinded approximately $90 million in appropriations from the LWCF for the four agencies. This rescission included more than $9 million intended for the habitats of endangered species. This amount was more than Congress had appropriated for that year. The president's rescission was possible only because the FWS had not yet spent some funds provided in the previous year.

The figures in table 8.3 also reveal rather large fluctuations in both presidential requests for the FWS as well as Congress's disregard for these requests. As an example, in the first budget for which the Carter administration had full responsibility, the FWS request to Congress more than tripled to nearly $27 million. This amount doubled in the administration's next budget request. In contrast, requests plunged sharply during the Reagan years. Despite the sharp reductions in the amounts requested, Congress was most generous in fiscal years 1984 and 1985. In the earlier year, no money was requested from the LWCF for the FWS. In response, Congress appropriated $27.4 million for endangered species, more than the cumulative appropriations for the previous four years. Though the Reagan administration wanted to reduce spending on domestic programs, including implementation of the endangered species program, the "zero-request" approach did not have the intended effect.

The wide swings in the amounts requested and appropriated not only undermine any notion that public budgeting is incremental, but also confound FWS efforts to develop long-range acquisition plans. The development of five-year plans was once an annual exercise. The Reagan administration's decision not to request acquisition funds largely negated the value of such plans, and they are now rarely pre-

Table 8.3   Requests and Appropriations from the Land and Water
Conservation Fund for the FWS, FY 1975–1991

Fiscal Year	Request to Congress		% of Total Request for En-dangered Species	Congressional Appropriations for Endangered Species	
	FWS Total (in thou-sands)	Endangered Species (in thou-sands)		Amount (in thousands)	% of Request
1975	$ 8,574	$ 1,300	15.3	$ 1,294	99.5
1976	8,500	6,900	81.2	6,400	92.8
Transition quarter	3,700	2,000	54.0	1,500	75.0
1977	8,495	4,500	53.0	14,900	331.1
1978	26,978	8,600	31.9	10,057	117.9
1979	53,710	27,840	51.8	18,115	65.1
1980	25,310	3,850	15.2	0	0
1981[a]	25,948	8,320	32.1	7,658	92.0
	(Carter) 11,420	−9,292	—	—	—
	(Reagan)				
1982	35,399	11,299	31.9	—	—
	(Carter) 0	0	—	1,967	—
	(Reagan)				
1983	1,567	998	63.7	9,853	987.3
1984	0	0	—	27,400	—
1985	29,000	14,550	50.2	31,036	213.3
1986	0[b]	0	—	16,648	—
1987	0[b]	0	—	15,100	—
1988	0[b]	0	—	13,980	—
1989	0[b]	0	—	16,535	—
1990	51,415	15,203	29.6	13,560	89.2
1991	56,700	7,300	12.9	NA	NA

Sources: Hearings and reports of the Senate and House Committees on Appropria-tions; Defenders of Wildlife, *Saving Endangered Species: Implementation of the Endangered Species Act in 1984* (Washington: Defenders of Wildlife, 1985); and FWS, Division of Realty.
[a]For fiscal year 1981, Congress appropriated $7.658 million in Land and Water Con-servation Funds for endangered species. During the first year of the Reagan admin-istration, $9.292 million was rescinded. The amount rescinded included the entire appropriation from fiscal year 1981 plus $1.633 million in unspent funds from the previous fiscal year.
[b]In fiscal years 1986, 1987, 1988, and 1989, the FWS requested $1.5, $1.5, $1.639, and $1.874 million, respectively, for acquisition management, but nothing for land acquisition.

pared. Ironically, the decision not to produce acquisition plans came at a time when Congress was providing more than the FWS anticipated spending to acquire habitats.

When the FWS purchases land for the protection of listed species, the land becomes part of the National Wildlife Refuge System. Access to and development within refuges is subject to some control. Consequently, the acquisition of land is often a far more attractive alternative than is designation of a critical habitat. With the latter, private activities can be completely unrestrained, and listed species can be subject to considerable stress due to these activities. In spite of the desirability of leasing or purchasing land or property rights for listed species, it is frequently a costly endeavor. Moreover, most endangered and threatened species could benefit from improved protection of their habitats, but money is unlikely to be available to provide this protection, at least for the acquisition of their habitats. Given limited resources and high demand for the resources that are available, how can they best be spent? Should $1 million be used to acquire habitat for a single high-profile species or for several obscure and relatively unknown species?

According to the FWS, it focuses its funds for acquisition on "only the most crucial needs."[40] Priority in the allocation of money from the LWCF is supposed to be given to species that face a high threat and that have a high potential for recovery. Similarly, species are supposed to have priority over subspecies. Other considerations include: "(1) relative costs, (2) land availability, and (3) whether the species' range has been sufficiently defined to enable FWS to determine what lands must be acquired to preserve them."[41] These preferences must be tempered by the realization that the government must also encounter willing private sellers. In their absence, use of the criteria becomes more difficult.

The FWS has indicated what criteria should affect its acquisitions, but the data in table 8.4 reveal that the Service does not always rely on its primary criteria, high threat and high recovery potential, to determine its LWCF expenditures. The table lists the fifteen species or habitats on which the FWS spent the most LWCF money since 1967, the first year in which habitats for listed species could be acquired. The degree of threat facing each species is also indicated at the approximate time the largest single expenditure was made.

The data in the table are instructive in at least several ways. First, and perhaps most noticeable, considerable amounts of money have

Table 8.4   The Fifteen Most Expensive FWS Acquisitions, Fiscal Years
1967–1989

Habitat	Sub-species?	Expenditures (in thousands of dollars)[a]	Degree of Threat[b]	Recovery Potential[b]
Mississippi sandhill crane	Yes	$20,576	High	Low
Key deer	Yes	16,779	High	Low
Whooping crane	No	16,000	High	High
American crocodile	No	14,934	High	High
Hawaiian water birds[c]	Yes	13,721	Low to Moderate	High
Florida panther	Yes	10,485	High	Low
Masked bobwhite	Yes	10,000	High	High
Coachella Valley fringe-toed lizard	No	9,389	Moderate	High
Light-footed clapper rail[c]	Yes	8,255	Moderate	High
Hawaiian forest birds[c]	Yes	7,065	Moderate to High	Low to High
Ash Meadows[c]		6,570		
California condor	No	6,612	High	Low
Bald eagle	Yes	4,994	Low	Low
Attwater's greater prairie chicken	Yes	4,844	Medium	High
Columbian white-tailed deer	Yes	3,651	Low	High

Sources: GAO, *Endangered Species—A Controversial Issue Needing Resolution* (Washington, 1979), 111–13; GAO, *Obligations and Outlays from the Land and Water Conservation Fund* (Washington, 1986), 14–25; and FWS, Division of Realty.
[a]These amounts include some administrative expenses for land acquisition or acquisition-related expenses such as survey and title work.
[b]Indications of the degree of threat and the recovery potential are based on FWS assessments in 1978, 1983, and 1986. The assessment provided for each species or habitat is the one chronologically closest to the year in which the single largest expenditure was made.
[c]The habitat provides protection for more than one endangered or threatened species.

been devoted to lease or purchase habitats for subspecies facing a low or medium threat and having doubtful prospects for recovery. Second, birds account for nine of the fifteen largest expenditures. This is in contrast to the situation with critical habitats for birds, which have rarely been designated in recent years. Third, about 40 percent of the largest acquisitions are in two states, Florida and California, noted for their high growth rates. Such growth typically provides an explanation for the decline of habitats. Fourth, listed plant species are rarely the intended beneficiaries of major acquisition programs. With the exception of the Ash Meadows area in Nevada, large acquisition programs favor birds, mammals, and fish.

Fifth, acquisition of habitat for at least one species, the bald eagle, is most likely due to its symbolic importance rather than for "crucial" biological reasons. Depending on its location, the species is classified as both endangered and threatened. Whatever its location, the FWS has consistently judged that the eagles face only a low threat. In one assessment released in 1978, the bald eagle was said to be in no immediate danger.[42] Having made this assessment, the FWS then spent money in seven of the next eight years to acquire habitat for threatened (rather than endangered) bald eagles in Oregon.

One should also remember that Congress makes an independent assessment of the distribution of funds from the LWCF. The FWS provides congressional appropriation committees with an annual priority list for acquisitions, but Congress is not bound to follow the list when appropriations are apportioned. For fiscal year 1990, as an example, the FWS requested just over $15 million to acquire habitats for endangered species. The congressional response was to provide about two-thirds of the request as well as another $3.5 million to acquire habitats for which the FWS had not requested support.

Regardless of the amount of money spent, leasing or purchasing all or part of a habitat cannot guarantee a species' survival or even its recovery. For several species identified in table 8.4, their situation has worsened rather than improved since acquisition of their habitats began. These include the key deer, the masked bobwhite, the light-footed clapper rail, and the California condor.

The experience with land acquisition for the dusky seaside sparrow also reinforces an unfortunate lesson. Starting in 1972, Interior began the eventual acquisition of over six thousand acres of habitat on Florida's east coast for the sparrow, a species with a low recovery potential. Few sparrows survived, despite the efforts. A survey of the bird's habitat conducted in 1979 found only twelve birds. All were males. By mid-1987 the FWS acknowledged that the species was extinct after the last male died in captivity.[43] In short, the commitment of any resource to a species with low prospects for recovery can be risky. The odds are against a favorable outcome except under the most fortunate circumstances. The seaside sparrows did not have this benefit; their habitat was small, and it encompassed part of the Kennedy Space Center. To reduce the mosquito population at the center, the sparrow's habitat was flooded, and this destroyed most nests.

However beneficial acquisition of habitats may be, it does not always provide a desired solution. Many species are so depleted that only a succession of biological miracles can ensure their recovery. Some habitats are not available for sale or lease while activities outside of acquired habitats often increase threats within the habitat. At other times the costs of habitats exceeds what Congress is willing to provide. Even in instances in which funds are available, the administrative and legislative processes associated with the approval of most acquisitions are time consuming. The FWS once estimated that "hundreds of discrete steps" are involved in the process.[44] The necessary evaluations and approvals can take more than six years before a habitat is actually acquired. For some species, such delays are intolerable, and emergency acquisitions can be handled in a more expeditious manner.

Finally, the appropriateness of the government's entire land acquisition program is subject to debate. President Reagan's Commission on Americans Outdoors recommended in 1987 that the LWCF be replaced with a new trust fund providing a minimum of $1 billion a year. This money would be dedicated to protecting open spaces for future generations.[45] Even before release of the commission's report, it faced substantial criticism from officials in the Reagan administration. Among their concerns, these officials complained that the government's purchase of additional land would increase federal spending, interfere with local economic development, and be interpreted as federal land-use planning.

The criticisms were not limited to those from public officials. One interest group sued Interior on behalf of landowners who opposed the commission's recommendations.[46] On the advice of the Department of Justice, Interior initially refused to release the final report. Even the commission's fifteen members were unable to obtain copies! At the least, such outcomes and strong opposition from many Republicans in Congress to the idea of more federal land purchases do not suggest a promising future for efforts to increase significantly acquisition of habitats for listed species.

Preventing Jeopardy to Listed Species and Their Habitats

Primarily as a result of the controversy involving the snail darter and the Tellico Dam, considerable concern existed that the FWS would henceforth have the legal authority to delay and possibly halt any federal project that adversely affects listed species or their habitats. The Service's projections of the expected number of consultations did little to mitigate these concerns. During congressional hearings in 1978, for example, the Service estimated that Section 7 requirements would produce nearly 20,000 consultations for the fiscal year beginning in October 1978. The Service's aggressive posturing in the mid-1970s also suggested, at least to some people, that many consultations would lead to jeopardy opinions. Such opinions, as noted earlier, indicate that a proposed federal project is "likely to jeopardize the continued existence" of a threatened or endangered species or result "in the destruction or adverse modification of habitat" of such species.[47]

Whatever fears might have existed, they have been largely unrealized. Only in 1989 did the number of consultations approach the estimates made for the late 1970s, and most consultations are now conducted informally. These informal consultations are optional, tend to be brief, do not result in formal biological opinions about a project's possible impacts on listed species or their habitats, and do not result in jeopardy opinions. Such consultations can include telephone calls and even inquiries to determine whether any listed species are found in the vicinity of planned federal activities.

In contrast to the large number of informal consultations, relatively few formal consultations occur (see table 8.5). These consultations are required whenever a federal agency determines that any of its activities may affect any listed species or its critical habitat. This requirement can be waived in instances in which there has been a prior determination, perhaps through an informal consultation, that the planned action "is not likely to adversely affect" a listed species or its habitat.

A formal consultation usually results in a biological opinion that discusses the anticipated effects of the action on listed species or their habitats. The opinion also indicates whether the action is or is not likely to jeopardize the continued existence of a listed species or result in the destruction or adverse modification of a critical habitat. These result in "jeopardy" and "no-jeopardy" opinions, respectively. Jeop-

Table 8.5   FWS Consultation Activities Under Section 7 of the
Endangered Species Act, Fiscal Years 1979–1989

Fiscal Year	Consultations		Jeopardy Opinions (in percentages)		Jeopardy Opinions (N)
	Informal	Formal	All Consultations	Formal Consultations	
1979	1,585	986	2.61	6.8	67
1980	2,374	707	1.75	7.6	54
1981	3,535	504	.72	5.8	29
1982	4,321	341	.48	6.7	23
1983	5,305	283	.59	11.7	33
1984	8,165	298	.27	7.7	23
1985	8,860	409	.40	9.0	37
1986	10,504	421	.48	12.4	52
1987	13,340	484	NA	NA	NA
1988	16,942	536	NA	NA	NA
1989	21,516	639	NA	NA	NA

Source: FWS, Division of Endangered Species and Habitat Conservation.
Note: Figures for fiscal year 1989 are preliminary estimates.
NA = Not available

ardy opinions must include a discussion of reasonable and prudent alternatives to the action that will avoid jeopardy, unless such alternatives are not available. If the action agency receives a no-jeopardy opinion or adopts the recommended alternatives contained in the jeopardy opinion, the agency is in compliance with the requirements of Section 7.

Jeopardy opinions are the most contentious since these are the ones that invite lawsuits or construction delays. Occasionally, a jeopardy opinion can even cause an action agency to terminate a project. It is important to emphasize again, however, that the action agency makes the determination of what to do with its project once it has received a jeopardy opinion. As the figures in table 8.5 indicate, however, the FWS rarely issues jeopardy opinions. Few listed species or their habitats cause problems for federal agencies, at least when measured in terms of the number of jeopardy opinions issued since 1979.

Windy Gap Opinions

To some extent the number of jeopardy opinions is misleading, particularly when one examines how the FWS has responded to some federal agencies or to applicants for federal licenses or permits. Whenever controversy erupts over the implementation of Section 7, the chances are better than average that a water development project is involved. This was the case with the snail darter, the protection of which led to court battles, congressional intervention, and, eventually, to changes in the Endangered Species Act. Many other controversies have similarly focused on water-related federal projects, but few have been as long-standing or as acrimonious as those involving water in the West.

For those familiar with western politics, there is an immediate recognition of water as a highly charged political issue. Disputes over allocation of the region's water have produced more than a dozen landmark Supreme Court decisions, long-standing disputes among the region's states, and even a threat from Mexico to sue the United States before the International Court of Justice. Water sustains agriculturally based economies in the West, and the availability of water has allowed the region's rapid population growth of recent decades. No better example of this exists than with the Colorado River. As Marc Reisner's graphic description indicates: "The river system provides over half the water of greater Los Angeles, San Diego, and Phoenix; it grows much of America's domestic production of fresh winter vegetables; it illuminates the neon city of Las Vegas. . . . It also has more people, more industry, and a more significant economy dependent on it than any comparable river in the world."[48]

Despite the services the river provides, it does not satisfy existing demands. The Colorado River once flowed unhindered through Colorado, Utah, Arizona, Nevada, California, and Mexico before entering the Gulf of California. Today, the Colorado's water is dammed, diverted, channelized, and withdrawn at scores of locations and "is so used up on its way to the sea that only a burbling trickle reaches its dried-up delta at the head of the Gulf of California, and then only in wet years." In short, the demands placed on the Colorado River and its tributaries help to explain why Reisner calls it "the most legislated, most debated, and most litigated river in the entire world."[49]

With the excessive demands placed on western waters, many Westerners are readily amenable to projects that offer the promise of increased water supplies. More often than not, these projects have some federal component or connection, thus requiring consultation under Section 7 when listed species are present. Agencies that offer such promises, like the BOR and the Corps of Engineers, thus tend to be popular, and their projects welcome. In contrast, agencies that attempt to limit or constrain water consumption on behalf of endangered species are likely to face hostility. Such has been the case for the FWS in its efforts to protect endangered and threatened fishes in the West. Some of these species are found only in the Colorado River basin and include the bonytail chub, the humpback chub, and the Colorado River squawfish. The latter species are charter members of the endangered species club; both were listed as endangered in 1967. The bonytail chub was listed in 1980. Of the three species, it is probably the closest to extinction.

In each instance the major explanation for the species' problems can be traced to human activities that have altered essential habitats without concern for the fish.[50] The Endangered Species Act's consultation requirements meant that such disregard would no longer be acceptable. Between 1977 and 1980, the FWS engaged in sixty-three formal consultations regarding projects to divert, store, or consume water in seventeen western states.[51] Most of these consultations produced no-jeopardy decisions. Nonetheless, the number of jeopardy opinions, which was twenty or 32 percent of the total, represented a far higher proportion than the national average. Most of the jeopardy opinions indicated that planned projects would adversely affect the continued existence of one or more listed fish species. This was particularly true for the jeopardy opinions issued for water-related projects in Utah and Colorado.

All these jeopardy opinions were widely criticized because of their perceived affects on planned projects. Indeed, some people protested that FWS implementation of the Endangered Species Act would halt all further water development projects.[52] Other critics were equally assertive in claiming that the jeopardy opinions had been based on either inadequate scientific data or even on speculation. Due to the uncertainty over the exact consequences of water depletions on some species, the FWS initially adopted a cautious approach that gave listed species the benefit of doubt. Until planned studies were concluded, as an example, the FWS once declared that proposed water withdrawals from two river systems in Utah would receive jeopardy opinions unless mitigating measures were implemented.[53] In other words, jeopardy opinions would be appropriate until the FWS finished the necessary research. Other biological opinions reached similar conclusions and emphasized the importance of considering the cumulative impact of many depletions rather than assessing each depletion in isolation.

As one representative from the Western States Water Council complained, however, jeopardy opinions "were being issued, or conservation measures required, based on only the assumption that . . . future depletions would adversely affect the recovery, if not the survival" of the fish species.[54] Other opponents of FWS actions claimed that the Service was usurping western water laws, the states' traditional responsibility to manage and allocate their own resources, and equitable apportionment decrees of the Supreme Court. Understandably, officials from these states complained to their congressmen, who in turn conveyed their irritation to Interior.

The FWS's solution was not altogether satisfactory to all participants. Faced with strong opposition to its jeopardy opinions, but still concerned with the protection of endangered species, the FWS attempted to reach a compromise in early 1981, in what became known as the Windy Gap approach. Windy Gap is a project in north-central Colorado that includes a dam and a scheme to divert water from the Colorado River. Based on its assessments and the anticipated consequences of the project, the FWS should have issued a jeopardy opinion; instead, the FWS attempted to avoid conflict. It agreed to issue a no-jeopardy opinion in exchange for the project sponsor's implementation of certain conservation measures that the Service required and payment of a one-time fee ($550,000) based on the amount of water to be depleted.[55] This fee would support further studies and habitat modifications for affected species. The FWS issued similar Windy Gap decisions in nearly four dozen subsequent opinions involving projects in the upper Colorado River basin. One result was that the number of jeopardy opinions involving projects in this basin plummeted. Indeed, for the next four years no water project in the basin received a jeopardy opinion.[56] In defending this approach, the FWS relied entirely on political considerations. The Service explained that the issuance of jeopardy opinions could have resulted in litigation, project delays, requests for exemptions from the Endangered Species Act, and perhaps changes in the law itself.[57] The FWS further justified its willingness to compromise by noting that the fees would provide research funds not otherwise available. In short, the FWS contended that the Windy Gap decisions represented reasonable and innovative solutions to conflict that allowed development while increasing the resources devoted to the protection of endangered fish.

Other observers were less charitable in their characterization of the approach. Some project sponsors argued that the data did not justify anything except no-jeopardy opinions. They also objected to what they considered to be extortion, that is, having to pay assessments in order to receive no-jeopardy opinions. Project sponsors claimed that they had no obligations to fund research projects on endangered species because their protection was a national rather than a regional goal. Through 1985, however, only two of the assessments exceeded $2 million. In contrast, seventeen assessments were for less than $15,000 each. One of these was only $70, a rather inconsequential amount. Moreover, the assessments were not due until after the water depletions were initiated, well after the commencement of construction or even after the completion of some projects.

The entire approach irritated many environmental groups, who complained that the FWS was far too accommodating to its sister agencies and prodevelopment interests. The FWS responded that it would use the sponsors' fees to conduct research on the affected species, but this meant that the Service would condone construction *before* the completion of its research and before identifying a project's likely consequences. In the case of Windy Gap, it was finished well before the research on the consequences of the project. As one critic pointed out, "[T]he FWS may have put itself in the position of being unable to alter the timing or magnitude of water releases by Windy Gap even if the studies indicated such changes were necessary." Furthermore, the FWS had failed to assess the effectiveness of the conservation measures it required project sponsors to implement.[58]

To summarize these concerns, endangered species were the beneficiaries of scientific uncertainty prior to the Windy Gap decisions. The Windy Gap process changed the beneficiaries so that construction activities could begin (and perhaps even be finished) without knowledge of their possible effects on endangered fishes. An assumption was made after 1981 that water projects were not likely to be detrimental to listed species' well-being as long as some research was undertaken and project sponsors initiated certain measures that the FWS deemed appropriate.

The FWS eventually ended the explicit use of Windy Gap opinions. In an attempt to contain possible conflict, however, the FWS established an Upper Colorado River Basin Coordinating Committee in mid-1984. Composed of representatives from state and federal agencies, the committee was charged with devising a method that would allow further water-related development in the basin while protecting endangered fish species.[59] The committee issued a draft recovery plan in late 1986 that attempted to reconcile competing interests. Briefly, the plan called for further research, propagation of the affected species, better management of existing water resources, and acquisition of water rights for fish- and wildlife-related uses. Whether these ideas will prove effective is uncertain; the plan anticipates few major changes in water usage or consumption. Federal agencies, such as the BOR, are expected to alter their operations of some reservoirs to release additional water at certain times of the year. Private water developers would still be required to pay one-time assessments based on expected annual depletions, just as the Windy Gap decisions required.

Although the Windy Gap approach marked a temporary period, the approach again reveals the Service's willingness to subordinate biological concerns to political expediency. Similar charges have been lodged against the FWS in regard to biological opinions issued for other projects. The FWS issued a jeopardy opinion involving the Concho water snake in 1986. After subsequent congressional pressure, the FWS reversed itself, despite the absence of any new data.[60]

Such practices are inappropriate and indefensible to hard-line advocates of strict interpretation of the Endangered Species Act. To these advocates, the needs of endangered species should not be subject to bargaining and compromise. Alternatively, such compromise may not be desirable, but it can represent an effective strategy that minimizes conflict with powerful private interests and that lessens the likelihood of further congressional backlash, as evidenced through reduced appropriations for or major changes in the Endangered Species Act. The FWS has too frequently found that it is not an effective and powerful administrative agency. For such an agency, its choices are limited; flexible interpretation of Section 7 that promotes some agreement is better than strict interpretation that leads to conflict and acrimony.

Responding to Jeopardy Opinions

The purpose of jeopardy opinions is not merely to inform federal agencies that their actions might be detrimental to a listed species or its habitat. The ultimate expectation is that an opinion, which must include reasonable and prudent alternatives to the proposed action, will cause an agency to alter its plans so that listed species or their habitats will not be harmed. Despite this expectation, the FWS has not effectively monitored projects to determine whether their sponsors have responded to the Service's recommendations. In fact, when one Interior official, Ronald Lambertson, was asked about this in late 1981, he observed that the FWS did not "have a mechanism to go back and monitor the project; we are putting that in place."[61] An examination of FWS data on consultations supports this observation. A few months after this statement was made, the OES released an analysis of jeopardy opinions issued in fiscal years 1979, 1980, and 1981. For slightly over 40 percent of the 178 instances in which the consultation process had been concluded with the issuance of a jeopardy opinion, the FWS did not know how the action agency had responded to the Service's recommendations.[62] In each instance, however, the action agency had proceeded with its project or had granted a permit or a license to a private applicant. More recent figures, for fiscal years 1982, 1983, and 1984, indicate a much improved situation. In only 12.5 percent of the jeopardy opinions issued in this period was the FWS unaware of the action agency's response to the recommendations.

These figures suggest substantial improvement in monitoring capabilities, but a more detailed analysis reveals this to be a premature conclusion. According to FWS data, other federal agencies canceled or withdrew projects in nine instances in which jeopardy opinions had been issued during fiscal years 1982–1984.[63] In only two of these instances, however, was concern for endangered species the primary or sole cause of the cancellation. In other words, the OES concluded that its jeopardy opinions had played an influential role in these two decisions; when asked about this, however, the action agency, the Corps of Engineers in both instances, reported a different outcome.

In the first episode, the FWS informed the Corps that plans to close a dike at the Oakland International Airport in California would jeopardize the continued existence of the California least tern. The FWS reported in 1985 that the Corps had refused to issue a permit for the project for this reason. In fact, the project applicant had already revised the original proposal to incorporate the alternatives the FWS had recommended. The Service's field office in Sacramento had rescinded the jeopardy opinion, and the Corps issued a permit in January 1983. In the second instance, the FWS issued a jeopardy opinion to the Corps in response to a private applicant's plans to plow wetlands for agricultural production. The Service concluded that plowing would destroy part of the critical habitat of the Cape Sable seaside sparrow. Despite this finding, the Corps did not believe that it had received a jeopardy opinion; nonetheless, it initially denied a permit to the applicant, but the applicant subsequently filed an amended permit application.[64] In short, the OES belief that jeopardy opinions had caused cancellation of both projects was incorrect.

Of far greater concern are the instances in which the FWS issued jeopardy opinions, the action agency rejected the alternatives, and the activity proceeded. In each such case, the FWS was aware that the alternatives had been rejected. Accordingly, though its monitoring capabilities were without problem, its ability to persuade its sister agencies to alter or cancel their plans further demonstrates the absence of effective sanctions in the Endangered Species Act as well as the Service's limited powers of persuasion. This is especially the case since one of the recalcitrant agencies was none other than the EPA, an agency normally sympathetic to environmental values.

These environmental values were at issue in EPA registration of pesticides. Such registration, or licensing, is contingent on pesticide manufacturers' willingness to comply with the conditions, standards, or restrictions that the EPA reasonably requires. In order to determine what conditions, if any, should be imposed to protect endangered species, the Ecological Effects Branch of the EPA conducts or reviews risk assessments on pesticides to be registered. A finding that a pesticide can harm a listed species requires the EPA to initiate a formal consultation with the FWS.

Thirty-nine such consultations occurred between April 1980 and November 1984. Of these, twenty-seven led to jeopardy opinions.[65] Most of these jeopardy opinions explicitly identified likely threats to listed species if they were to be exposed to the pesticide in question. Furthermore, in all but one of the opinions, the FWS included recommendations on how jeopardy might be mitigated, most often through restrictions or prohibitions on use of the pesticide in certain counties. Internal EPA guidelines prohibit the registration of pesticides that jeopardize a species' existence, so the expectation was that the EPA would readily accept the FWS recommendations. In fact, however, the EPA did not restrict the use of even one of the pesticides that had received a jeopardy opinion. In at least six consultations that resulted in jeopardy opinions, the EPA initiated the consultation process, but then issued a registration permit before receiving a biological opinion from the FWS.

In the case of chlorpyrifos, an insecticide used to protect field crops, its manufacturer requested registration in early 1980. Consultation with the FWS was requested, but the request was soon withdrawn after the Ecological Effects Branch recommended against registering the pesticide. The branch was concerned about what it called "the projected unreasonable adverse effects to nontarget organisms" that exposure to chlorpyrifos would entail. The manufacturer insisted on seeking registration approval, so the FWS was once again asked to provide a biological opinion. Six weeks *before* the biological opinion was issued the Registration Division of the EPA approved the pesticide's registration. When the FWS did provide an opinion, it noted that use of the pesticide might jeopardize over 100 listed species and their critical habitats. The species ranged from fish and birds to insects, mammals, reptiles, and amphibians. To prevent harm to these species, the FWS suggested buffer zones and the use of geographic restrictions and certain methods of insecticide application. The Registration Division rejected all of the recommendations.

For metolachlor, a herbicide used to control weeds, both the Ecological Effects Branch and the FWS concluded that its use would jeopardize nearly a dozen listed species. Jeopardy could be avoided, however, if the herbicide was not used in certain counties that the FWS identified. The EPA could have required metolachlor's manufacturer to list these counties on the product's label. Instead, when the agency registered the herbicide, it chose not to require labeling that reflected concern for endangered species. As one EPA official explained, the Registration Division does not like too much information on labels.[66] Information on the affected species was judged to be too long, so it was not included on containers of metolachlor.

The experiences with chlorpyrifos and metolachlor are illustrative of disagreement within the EPA. For both pesticides, plus at least seven others, the Ecological Effects Branch agreed with the jeopardy opinions and their recommended alternatives. In all nine instances, however, the Registration Division rejected the alternatives and allowed registration, either before or after the consultation was concluded.

The EPA evidenced still further insensitivity because it did not have a system to monitor how the actual use of pesticides affects endangered species. The Center for Environmental Education noted that these "effects could be the result of oversight during the regulation process, the dynamic nature of endangered species, or an unforeseen property of the pesticide."[67] Although these are the possibilities, the center found that the EPA had no effective mechanism for reporting unintended pesticide poisoning. Though it occasionally received information indicating that pesticides had killed some individual members of an endangered species, the agency had not taken any action to investigate the deaths. The EPA had similarly not taken any enforcement actions after the pesticide-related poisoning of listed species.

To ensure attention to the effects of pesticides on endangered species, Congress subsequently required the EPA to initiate an extensive educational campaign among agricultural users of pesticides.[68] In addition to this requirement, the Endangered Species Act Amendments of 1988 also obligate the EPA to examine how its pesticide-labeling program can be improved in order to protect listed species as well as to minimize the impacts on pesticide users.

Conclusions

It is not possible to generalize from the EPA's experience with pesticides and jeopardy opinions. At the least, however, the experience provides cause for concern about how other agencies respond to jeopardy opinions and the requirements of Section 7. The EPA has its own standard operating procedures regarding pesticides and consultation as well as a legal requirement in the Federal Insecticide, Fungicide, and Rodenticide Act to monitor the consequences of pesticide use. Both the procedures and the law were routinely ignored, in part because EPA management led its employees to believe that compliance with Section 7 was a low priority.[69] If this was the attitude in an agency with a mandate to protect the environment and one that is well staffed with environmental scientists and conscientious professionals (at least in the Ecological Effects Branch), what are the prospects that other federal agencies with no such mandate and with many fewer qualified scientists will be more concerned about endangered species than is the EPA?

Similarly, the experience and other data presented in this chapter suggest that Section 7 is not an adequate mechanism for protecting endangered species. Action agencies are expected to initiate consultations once they determine that their projects (or projects for which they must issue permits or licenses) may jeopardize listed species or their critical habitats. This expectation is based on the assumption that these agencies and all their regional, district, and field offices are capable of making such determinations, that the necessary data are available to make an informed judgment, and that these offices are willing to provide the information to the FWS. All three assumptions are questionable. Government agencies exhibit a natural bureaucratic tendency to delay the release of information or to withhold it altogether from other agencies because this is how power and influence are exercised.[70] In order for federal agencies to comply effectively with Section 7, they must know where endangered species are located and the likely consequences of their planned activities on these species and their habitats. The likelihood of this occurrence is problematic.

On the one hand, many species migrate, so their habitats vary; old habitats are abandoned while new ones are occupied or discovered. The decision not to designate critical habitats for most species also frustrates federal agencies' efforts to comply with Section 7. At least this was once the position of Interior. Robert L. Herbst, one of the department's assistant secretaries, emphasized that when critical habitats are not designated, federal agencies do "not have all of the data useful in making decisions as to whether their programs or actions are compatible with the requirements of section 7."[71]

On the other hand, thoroughly and accurately assessing the future consequences of today's actions is impossible. The FWS cannot do this, so is it reasonable or appropriate to assume that other federal agencies can? Although the evidence is fragmentary, there is good reason to believe that many agencies, unintentionally or otherwise, overlook listed species that their activities affect.[72] Having the action agency, rather than the FWS, conduct the initial biological assessment provides still another reason to conclude that some agencies are unaware of or unable to determine the biological effects of their projects. From the perspective of listed species, the FWS should also be expected to issue biological opinions without concern for the reaction of the affected agencies or permit applicants. This has not always happened, and nonscientific variables seem to enter the Service's decisionmaking calculus frequently enough to justify concern.

One way to improve the effectiveness of the consultation process would be to incorporate either sanctions for noncompliance or incentives for compliance. Much recent research has shown that one or the other approach is almost always essential, particularly when governments attempt to protect common property resources such as endangered species. In contrast, exhortation in the form of requests for compliance is rarely productive.[73] The choice of governing instruments, ranging from sanctions to incentives to exhortation, usually reflects a government's preferences about the relative importance of different goals. The most stringent penalties and the most desirable incentives are thus reserved for a society's most important goals. The absence of effective sanctions or incentives signals that a society attaches less importance to a goal.

This discussion is pertinent to Section 7. The FWS is without authority to decide whether another agency's project should proceed in the face of anticipated harm to a listed species. This responsibility is left to the action agency. The FWS can apply whatever bureaucratic pressures it has in its administrative arsenal, but it cannot force an action agency to alter or terminate a project because of concerns about endangered species. Private organizations or individuals can sue the action agency on behalf of the affected species, but this usually represents only a hypothetical sanction because it is used so infrequently. Throughout the four-and-a-half years that the EPA routinely disregarded jeopardy opinions on pesticides, it faced no lawsuits and no penalties for its actions. The EPA thus operated with total impunity even though the FWS, Congress, and several national environmental organizations were aware of what was happening.[74] For most jeopardy opinions, however, it is probably far more difficult to monitor agency response. The FWS has decentralized the responsibility for issuing most biological opinions to its regional offices. In turn, these offices usually delegate the task to field offices, which are located throughout the country. Copies of jeopardy opinions eventually reach FWS headquarters, but this does not significantly enhance environmental organizations' ability to keep track of individual projects or to determine whether agencies are responding to the reasonable and prudent alternatives included in jeopardy opinions. The problem might be partially mitigated if the FWS required agencies to justify their disregard of jeopardy opinions, but the Service has declined to do so.[75] The Service again believes that it is without legal authority to require such justifications.

Several other aspects of the consultation process deserve brief mention. The first issue is that Section 7 is entirely silent in regard to private actions that have no ties to federal agencies. The appropriateness of this choice is debatable, particularly if one believes that listed species deserve protection regardless of where they are found.

The second issue involves the geographic scope of Section 7's consultation requirements. Many endangered species live outside the United States, and federal activities sometimes harm these species. The consultation regulations issued in 1978 required every federal agency to ensure that its activities in the United States, "upon the high seas, and in foreign countries" not jeopardize the continued existence of such species.[76] This extraterritorial requirement was included in the regulations without any apparent disagreement. In the proposed and final consultation regulations issued in 1983 and 1986, respectively, the obligation to consult in regard to species in foreign nations was eliminated in the belief that consultation might infringe on the sovereignty of these nations.[77] The Defenders of Wildlife and several other environmental organizations successfully challenged the FWS interpretation of the law.[78] A federal district court ruled in early 1989 that federal agencies must consult with the FWS for overseas projects, but the court also delayed execution of its order pending the outcome of an appeal of the decision by the Department of Justice.

Last, one can also question the burden-of-proof requirements in Section 7. In a sense, the section requires that an action agency and the FWS conduct an assessment of anticipated risks each time a project has the potential to harm an endangered species. A statement from the Science Council of Canada captures the assumption implicit in this approach:

> If an inquiry looks upon its work as an assessment of risk, it has already made a value judgment—it assumes that a project . . . should be permitted *unless* significant problems can be identified that call for reconsideration. To those seeking a review of government policies, such an approach is often unsatisfactory because it may preclude a fair discussion of whether a particular project . . . would be appropriate or desirable.[79]

An alternative would require project advocates to demonstrate that their activities are not harmful to endangered or threatened species. This approach would place far greater demands on federal agencies than now exist, but somewhat comparable requirements are imposed on drug manufacturers. These manufacturers must demonstrate to the satisfaction of the Food and Drug Administration that their products are safe and effective before they can be marketed. The procedures for demonstrating safety and efficacy are quite rigorous, but they do appear to be reasonably effective in protecting the American public. Opponents of the approach can point to similar burden-of-proof requirements placed on pesticide manufacturers. Unfortunately, as discussed above, these requirements were not useful in protecting endangered species, at least in the 1980s.

As this and the previous three chapters have demonstrated, the task of listing and protecting endangered species is an enormously complex one. In fact, however, the discussion so far has actually understated this complexity. The FWS has primary responsibility for the implementation of the Endangered Species Act, but it does not have sole responsibility. It is, therefore, important to consider how the FWS interacts with its fellow implementors.

9   Divided Jurisdictional Responsibilities:
Seeking Effective Cooperation

Protecting endangered species involves more than a single agency. All states have some listed species, and not all are found on land that the FWS owns or manages. Many birds and mammals are migratory, other species never leave the oceans, and still others divide their time between land and water. Similarly, measures to protect endangered species are implemented with some recognition of hunting, recreational activities, and economic development. In short, scores of government agencies have some interest, albeit diverse, in plants and animals. For a smaller number of agencies, this interest is quite intense, either because of the nature of the species (e.g., many animals serve as targets for hunters), the nature of the agency's mission (e.g., the development of commercial fisheries), or because of an interest that predates the involvement of the federal government (e.g., a state's concern for resident wildlife).

In drafting laws to protect endangered species, Congress could have put aside these interests and assigned sole responsibility for implementation to a single agency. Alternatively, Congress could have divided responsibility and recognized the existence of diverse interests of several agencies as well as the states. Congress chose the latter course. Accordingly, this chapter examines some of the consequences of this choice and focuses on how divided responsibility affects both the listing process and the designation of critical habitats. Before doing so, however, it is first useful to consider some possible organizational consequences of single-agency responsibility versus responsibility that is shared among several organizations and more than one level of government.

Sole Versus Divided Responsibility

Relying on a single agency to implement a policy has at least some readily apparent advantages. On the one hand, the locus of decision-making is obviously within one agency and this provides some economies of scale. One, and not several agencies possess the necessary expertise. Having a single agency in charge of a program also allows that agency to serve as a comprehensive clearinghouse for information and advice. When problems develop, one can quickly identify the agency with responsibility and the likely authority to address the problems.

On the other hand, few administrative agencies, regardless of their goals or mission, like to share responsibility for their key functions. Instead, agencies prefer to have sole responsibility, to have decision-making within their control, and to have independent authority to craft rules and regulations, to ascertain relative priorities, and to focus on problems that they and their constituents deem important. Administrative agencies are sensitive to intrusions into their interior zone, the area in which they are the major determinants of policy. In many ways this sensitivity is appropriate. Many studies of the implementation of public policies identify interagency relations as a key explanatory variable.[1] To the extent that one agency must coordinate its activities with another, the prospects for disagreement will increase, even when both agencies are on friendly terms and share similar goals. Establishing effective working relations is rarely an easy task.

Relations with other agencies may not always be friendly, however, and maintaining effective ties is likely to involve bargaining, negotiation, and the commitment of scarce organizational resources. For these reasons alone, agencies are frequently reluctant to share their policy responsibilities. In addition, the ability to develop effective interorganizational relations presumes that sister agencies share collective goals or are at least willing to compromise their objectives to accommodate competing interests. From any agency's perspective, the least desirable arrangement is one in which its sister agencies express indifference or perhaps hostility and when coordination is statutorily mandated. This situation is likely to produce acrimonious relations, ineffective shared responsibility, and perhaps even stalemate.

Whatever the supposed advantages of single-agency responsibility for a program, there are advocates of divided responsibility. President Franklin D. Roosevelt liked to spread responsibility for the same tasks among different agencies. He wanted to encourage competition, thereby producing more effective policies.[2] Competition among agencies can also increase the attention devoted to a problem, stimulate innovation, and perhaps broaden perspectives so that parochial interests do not prevail at the expense of the public interest. Other observers contend that redundancy enhances reliability.[3] The elimination of overlapping jurisdictions reduces a governmental system's effectiveness because the failure of one part could lead to failure in the entire system, or so the argument goes.

However one views the virtues of sole or shared responsibility, one must also remember that rational decisionmaking processes are not likely to provide the best explanation for decisions about jurisdiction. Once an agency has responsibility for a policy area, it is unlikely that the agency will voluntarily relinquish the responsibility. To do so risks a loss of power, resources, discretion, clientele, and control over the development and implementation of policy. With such consequences, opposition to a change in jurisdiction is likely to be intense, and the burden of proof falls on the advocates of change. This burden will be extraordinarily difficult to meet when the opponents of change include the states, whose interests are quintessentially represented in Congress. In other words, existing patterns of responsibility are not easily changed, and this probably provides the best explanation for the role that the states and the NMFS play in the listing of species and in the designation of critical habitats.

Interorganizational Coordination: The FWS and the NMFS

As noted in chapter 3, the Fish and Wildlife Act of 1956 created two organizational entities within the FWS, namely the Bureau of Commercial Fisheries (BCF) and the Bureau of Sport Fisheries. This was an unhappy bureaucratic marriage, and it eventually led to the transfer of the former to the Department of Commerce in 1970. With the transfer also came a new name. The BCF became the National Marine Fisheries Service (NMFS), which is now part of the Commerce Department's National Oceanic and Atmospheric Administration.[4] With this reorganization, some decision had to be made about which department, Interior or Commerce, would be given management authority for marine mammals. Under the terms of the separation, Interior retained authority for manatees, walruses, sea otters, and polar bears.

Commerce and, in turn, the NMFS, received responsibility for seals, whales, dolphins, porpoises, and sea lions. When this division was made it had nothing to do with the protection of endangered species. Given the agencies' interests, abilities, and preferences, the division did seem to make some sense. No doubt, too, an element of compromise was reflected in the division of species.

This division provided the justification for subsequent congressional policymaking. In what was one of its more controversial provisions, the Endangered Species Act of 1973 similarly divided responsibility for listing marine species between the two departments. For the species for which Commerce has responsibility, the 1973 law *authorizes* the secretary of commerce to determine whether any such species should be listed as endangered or threatened or have their status changed from threatened to endangered. Additionally, the law *allows* the secretary of commerce to recommend the delisting of such species or their reclassification from endangered to threatened. Interior retains formal authority to list or delist any of these species but cannot do so without "a prior favorable determination from the Secretary of Commerce."[5] This process was clearly a compromise between competing interests. Fearing that the listing activities of the FWS might adversely affect its policies and objectives, Commerce had once asserted that it should be allowed to veto any listing proposals involving marine species.[6]

Regardless of the nature of the compromise, many environmental groups and some Interior officials had opposed the split jurisdiction during congressional hearings. These opponents believed that Commerce tacitly favored exploitation. To support their position, these opponents pointed to Commerce's experiences with the protection of whales. In late 1970, when Interior still had responsibility for listing all marine species, the department's secretary, Walter J. Hickel, announced that eight species of whales would soon be listed as endangered. Three of these species, the fin, the sei, and the sperm whale, were subject to extensive international commercial harvesting and were critically endangered, at least according to the evidence Hickel presented. By classifying the species as endangered, the importation into the United States of all whale products would be prohibited.

Maurice H. Stans, the secretary of commerce, immediately protested what he termed a unilateral and scientifically indefensible action. On the one hand, Stans said there was little evidence that the fin, sei, or sperm whales were endangered. "On the other hand," he added, "the damage to industry from a ban on sperm oil would be severe." This kind of pressure had quite an apparent effect. The day after Hickel's announcement (but before publication in the *Federal Register*), his decision to list the whales was reversed, and he was fired.[7]

These decisions generated considerable adverse publicity, and the whales were soon listed as endangered. Commerce, exercising its authority, allowed U.S. citizens to capture and kill these whale species until the end of 1971, a year after they were listed.

Environmental groups believed that the whales' tale provided more than enough justification to deny Commerce any responsibility for protecting endangered species. Despite the groups' opposition, their efforts in 1973 were insufficient to overcome the existing division of responsibilities, the Nixon administration's preferences for the arrangement, and the likelihood that the congressional committees considering the legislative changes—the Senate Commerce Committee and the House Committee on Merchant Marine and Fisheries—were sympathetic to these preferences.

For those apprehensive about dividing responsibility for listing marine species, the events that followed vindicated their concern. The FWS proposed in late 1973 that two species of sea turtles, the green and the loggerhead, be listed as endangered. The two species had declined precipitously. The green turtles were believed to have less than ten thousand individuals from stocks that had once approached fifty million. Human activities, including shrimp trawling and incompatible ocean-front developments, the FWS stated, were victimizing the two species.[8] Based on past FWS efforts, once a species was proposed for listing, one could expect a final rule to follow in about four or five months.

In this instance, past experience provided no guide. On the same day that the turtles were proposed for listing, President Nixon signed the 1973 Endangered Species Act, thus dividing responsibility for marine species along the lines already indicated. Responsibility for marine mammals was specified, but not for sea turtles, which lay their eggs on beaches. The new law did not clarify which agency would handle turtles and jurisdictional ambiguity produced confusion and conflict.[9] Prior to the act's passage, when Commerce clearly did not have any responsibility for listing, the department had favored endangered status for the turtles. With the law's passage, however, the department's preferences soon changed. The result was years of delay and frequent battles over jurisdiction.

Commerce claimed responsibility for the turtles, but Interior resisted. Due to the uncertainty over which agency had responsibility for listing the turtles, the proposal to list them was withdrawn. Next, since the FWS and the NMFS did not agree, the directors of the two agencies began discussions about how they would share or divide administrative responsibility for the Endangered Species Act. These discussions produced a memorandum of understanding on responsibilities and listing procedures, which both directors signed in August 1974.[10] Although the memorandum settled some problems, others remained unsettled, including the most contentious. The directors agreed that until the turtle issue was resolved, they would jointly implement all actions regarding the turtles. Eventually the two agencies reached an apparent agreement, and they issued a joint proposal to list the green and loggerhead turtles as threatened, rather than as endangered as originally proposed. In addition, a third species, the Pacific ridley sea turtle, was also proposed as a threatened species.[11]

The FWS and the NMFS noted that the three species require protection because of overutilization, threats to their habitats, and the inadequacy of existing protective measures. However severe the turtles' problems, the proposal included provisions to allow mariculture operations to breed green turtles and the continued importation of turtle products for at least two years after the final rule was published. These provisions were intended to mollify importers and existing mariculture operators, but these interests were not appeased. The National Canners Association, for example, requested a public hearing to discuss what it considered to be the "gravity, complexity, and controversial nature" of the proposal. Not only did the NMFS agree to hold a public hearing, but it also chose to do something never before done in regard to a proposed listing. Though the FWS opposed the idea, the NMFS decided to prepare an environmental impact statement on the proposed rule, thereby adding still more time to the listing process. The draft statement, which was released in early 1976, confirmed that the turtles should be listed, but the statement neither settled the jurisdictional conflict nor added the species to anyone's list.

Still another year passed before the agencies could decide how to divide management responsibility for the turtles. In an agreement that satisfied neither agency, the FWS was given responsibility for the turtles when they are on land and the NMFS when they are in the water. It was expected that this agreement would allow the turtles to be listed almost immediately, but another year passed before their listing. Finally, in late July 1978, more than four years after the original proposal, the three turtle species were listed as threatened.[12] No exceptions were provided for mariculture operations.

The inability to reach a quick solution to the turtle dispute is not surprising to those familiar with interagency disputes over jurisdiction. The FWS believes that it should have at least some responsibility for all species at risk of extinction. The NMFS has a constituency that favors the use and capture of marine resources. Such commercial interests typically feel threatened by any effort to affect their livelihood; listing species can easily trigger that concern. To remain responsive to its constituents, the NMFS cannot ignore these commercial pressures. One result is that since 1973, when the NMFS first gained responsibility for certain marine species, the agency has not initiated any proposals to list species. All its listing actions have come in response to petitions or the threat of legal action. Other than the turtles discussed above, through mid-1989, the NMFS had listed but four other species, only one of which, the Guadalupe fur seal, is native to the United States. Two of the four listing actions occurred in 1985, in one instance more than six years after the NMFS received a petition from an environmental group to list the Gulf of California harbor porpoise.

A further indication of the NMFS's lack of concern for endangered species can be seen in its commitment of resources to the program. These resources have not matched congressional expectations. Occasionally, discussion about these resources has provided opportunities for humor, as the following dialogue between two members of Congress and a deputy director of the NMFS revealed in the mid-1970s:

> Congressman One: "Let me be sure I understand what you are saying. Your entire endangered species office is sitting in the front row, is he not?"
>
> NMFS Deputy Director: "Yes, Sir."
>
> Congressman One: "Does he have any staff working directly for him?"
>
> NMFS Official: "No, he does not. He does call upon other staff for assistance in clerical and stenographic work."
>
> Congressman Two: "Well, for purposes of the record I think the record should reflect what you . . . are saying is that you have one person in your agency who is handling the endangered species. . . ."[13]

The NMFS's resource situation did improve from this point, but its funding was vulnerable to the budget cuts of the Reagan administration. For fiscal year 1986, as an illustration, the administration requested a decrease of nearly 40 percent in expenditures for NMFS activities on what it calls protected species.[14]

Intergovernmental Coordination: The FWS and the States, a Case Study of Hawaii

The FWS and the NMFS are not the only actors involved with the protection of species. Both agencies must also contend with the states. Cooperation from the states is usually essential in protecting listed species, so it makes sense to incorporate state views into decisionmaking processes. In addition, however, substantial historical justification exists for the consideration of state perspectives. For many years before active federal involvement in the management of fish and wildlife, states had responsibility for these species. Indeed, as a result of a series of judicial decisions dating to the 1840s, the traditional presumption in the United States was that states had responsibility for resident wildlife. Many states jealously guarded this prerogative and opposed federal efforts to become involved in the management of resident as opposed to migratory species. Federal laws gradually eroded this presumption of state primacy, but endangered species laws still recognize the states' interests in resident wildlife, particularly when it comes to listing species. In fact, federal laws governing the protection of endangered species make clear that the secretaries of commerce and the interior must consult with affected states before listing species found within those states. This consultation is intended to incorporate the states' views into decisionmaking processes. Consultation can, however, also lead to negotiation and, occasionally, compromise because states that oppose a species' listing can expect to enlist sympathetic congressional involvement and, hence, effective pressure on the FWS. Moreover, state fish and wildlife officials outnumber their federal counterparts, so the latter often find themselves dependent on the cooperation and good graces of the former. This dependence can lead to responsiveness, as the FWS's experience with Hawaii suggests.

Among all American states none has as many species on the edge of extinction as does Hawaii, and no other state can claim as many recent extinctions. Since the time that Europeans first visited the

Hawaiian Islands about one-third of the state's endemic bird species have disappeared; of the remaining forty-four bird species found nowhere except Hawaii, twenty-nine are listed as endangered and one as threatened. Of these thirty avian species, one is presumed to be extinct while the known wild populations of three others are believed to be less than ten.[15] The state's only native mammals, the Hawaiian hoary bat and the Hawaiian monk seal, are endangered. Hawaii also has more extinct plant species and more plants in danger of extinction than any other state. When the Smithsonian Institution issued its list of plants believed to be endangered in the United States, Puerto Rico, and the Virgin Islands, over 45 percent were Hawaiian. The FWS assessment of the state's plants is similar. In a report published in 1980, the FWS noted that for nearly eight hundred Hawaiian plants, it has enough information either to justify listing or the probable appropriateness of listing.[16] Of the eight hundred, about 16 percent are believed to be extinct. The FWS further identified fifty-one American plants for which it has "persuasive evidence of extinction." Forty-five of these had once been found only in Hawaii.

Not only does Hawaii have a disproportionately large number of species in danger, but it also has a high number of species that are exceptionally close to extinction. Based on its analysis of the relative jeopardy of listed species, the FWS believes that for every five endangered species in Hawaii, three face "almost certain [extinction] in the immediate future because of a rapid population decline or habitat destruction." A 1989 assessment of Hawaii's endangered flora estimated that nearly one hundred of its native species are in such danger that they may perish within five years.[17] In short, for many of Hawaii's endemic species, life is a perilous venture.

While the loss of any species may be undesirable, it is particularly so for Hawaii's native species. Many of these species evolved in and became dependent on highly fragile ecosystems. The Hawaiian Islands are the most geographically isolated land mass in the world and, as one scientist, Alan D. Hart, has observed: "Isolation, time, and habitat diversity have produced a most distinctive biota. More than 97 percent of Hawaii's native flowering plants, nonmigratory land birds, insects, and land snails occur naturally nowhere else on earth."[18] For many species this isolation was an ecological boon. One entomologist, Wayne C. Gagné, has provided a graphic description of the islands' advantages: "No need . . . for an island plant to be poisonous or thorny, no need for the island bird to be eternally vigilant, no need for an island insect to guard against the voracious ant—those were all continental constraints."[19] Isolation provides a strength, but it also creates a high vulnerability to disruption. Islands typically offer protection from predators, but when they are introduced native species are limited in where they can go and how rapidly they can develop defense mechanisms.

When change occurs rapidly, previously isolated species become helpless victims. It has been just such change that has threatened so many Hawaiian species. In the millions of years prior to human settlement nature successfully introduced about 250 plant species, about 250 insect species, 16 different landbirds, and only 2 mammals to the islands. A new plant or insect species was introduced and successfully adapted to the Hawaiian environment on the average of about once every 75,000 years.[20] For landbirds and mammals, successful introductions occurred about once every 1.25 million and 10 million years, respectively. Polynesian and European immigrations over the last two thousand years have brought enormous increases in the number of exotic species and their rates of introduction. The number of wild mammal species is now more than ten times what it was before human settlement. Wild or feral mammals in the state now include dogs, cats, pigs, goats, sheep, and rabbits, all of which alter ecological stability. As an example, rats prey on the nests of many birds; to rid the islands of rats, mongooses were introduced, and they too prey voraciously on native birds.

New introductions have not been limited to mammals. Nearly two dozen exotic insect species find their way to Hawaii each year—a rate more than 1.5 million times as rapid as nature allowed prior to human settlement. The number of introduced plant species probably exceeds four thousand. Nearly all these introductions severely challenge and sometimes exceed the ability of native species to cope with the consequent changes.

Other human activities have had similar detrimental consequences for many of the state's endemic species. Tens of thousands of acres of virgin forest have been destroyed in order to plant exotic trees, sugarcane, and pineapple. To care for these and other crops, herbicides and pesticides are heavily used. One report from the state's Department of Agriculture noted that in the 1960s Hawaiian agricultural interests applied about ten times the amount per acre of herbicides and pesticides as was used on the mainland.[21] Local termite exterminators are said to use 500 to 1,000 times as much poison per application as do their counterparts in other states. All these activities reveal that human activities provide the single best explanation for the problems of endangerment in Hawaii.

The plight of Hawaii's species might lead one to expect that the state would be at the forefront of efforts to preserve its native species. In fact, however, this has not been the case. Development may be undesirable for many species, but it is this development that spurs and sustains much of the state's economic growth. Hawaii's permanent population is about one million, but the state's population density and growth rate are more than twice the national average. Millions of other people visit the islands each year as tourists, and demands for hotels, condominiums, housing developments, and recreational opportunities remain high.

Land development has been a high priority in the state since at least the early 1950s, and the political system has frequently encouraged and facilitated this development.[22] Some state officials believe that requirements for the protection of endangered species could jeopardize continued economic growth. One manifestation of this situation is that the number of Hawaiian species officially classified as endangered or threatened is far below the number one might expect given the severity of the threat and the number of species, especially plants and invertebrates, that are candidates for listing.

Of the fifty-two endemic Hawaiian species listed as either endangered or threatened through 1989, more than half were listed prior to 1971, when the requirements for listing were much less stringent than they are today. The first listing of a Hawaiian plant did not occur until 1978. The pace of plant listings increased after 1978, but fifteen years after the issuance of the Smithsonian Institution's report, only nineteen Hawaiian plants were on the endangered species list, and most of these were listed in the mid-1980s.

In several instances, the state has opposed the listing of species even though its objections were not well founded. As an illustration, the FWS proposed in 1980 that the 'Ewa Plains 'akoko be listed as endangered. Governor George Ariyoshi opposed the proposal because he believed the plant's listing might interfere with the Corps of Engineers' proposed development of a deep-draft harbor on Oahu, the state's most populous island. "The listing of this plant," declared the governor, "will have a severe and adverse economic impact on Hawaii."[23] As a second line of attack, the governor pointed out that the land's owner had transplanted some of the plants in 1977 and 1978, and some of them were still alive, thus suggesting that additional plants could be successfully transplanted. What the governor did not mention was that only 1 percent of the plants had survived the earlier transplant. The governor's opposition was further undermined when the Corps agreed that the 'akoko was in real danger, that its existence would not affect the deep-draft harbor, and that the species should be listed immediately. In regard to the alleged economic impact of the proposed listing, the FWS estimated that the cost of the necessary protective measures was unlikely to exceed $10,000.

The 'akoko was eventually listed as endangered, but still other species might have been proposed except for bureaucratic disagreements within Hawaii's Department of Land and Natural Resources (DLNR). There exists an intentional effort to frustrate or discourage the federal listing of species, at least according to some of its employees.[24] One department employee indicated that the state botanist had prepared a list of more than fifty plant species to be recommended to the FWS for listing. After the botanist's superiors reviewed the list, only five species remained on it, but none of these were recommended to the FWS. In other instances, departmental scientists were aware of the locations of candidate species but were then prohibited from informing the FWS. The department occasionally received petitions to place plants on the state's endangered species list, but the state botanist was not allowed to review the petitions.

The reluctance to list species is clearly unrelated to biological considerations, or so some DLNR employees believe. These employees point to low levels of political support for their activities, politicians' natural reluctance to arouse controversy and their belief that protecting species will retard development or limit what private landowners can do with their property, and district foresters' unwillingness to impede hunters or to assume additional management responsibilities.

This political climate appears to affect FWS activity as well. Most proposals to list species originate in the Service's field or regional offices, so the distribution of FWS personnel provides a measure of its willingness to list Hawaiian species. Hawaii has more plants that are candidates for listing than any other state, but in the 1980s the FWS assigned only one botanist to initiate listings in the state. Asked about this apparent inconsistency, one OES scientist answered that the FWS could easily justify a threefold increase in the number of botanists on its listing staff in Honolulu.

Although an increase in staff can be justified, the decision not to be more assertive probably reflects the reaction of the FWS to the political climate in Hawaii. This climate, according to an FWS official involved in the listing process, is hostile and "phenomenally antagonistic" to any kind of outside interference.[25] As this official noted, getting species listed in Hawaii is "like pulling teeth": the state government provides either poor or very mixed cooperation and it has "bitterly opposed" many proposals. In view of this resistance, the FWS's choices make political rather than biological sense. The FWS official summarized the situation in this way: "Why buy yourself all that trouble in a small, remote, isolated place? I think that probably governs a lot of our staffing decisions." Whatever the explanation, the result is that no Hawaiian species were listed as either endangered or threatened in 1987, 1988, or 1989.

For the Hawaiian species that do get listed, they tend to be on public land and be noncontroversial. Thus, when the FWS wanted to classify the Cooke's kokio as endangered, the state did not oppose the proposal. None of the plants are found in the wild; the species survives only in a botanical garden in Honolulu.

The kokio provides some evidence about the noncontroversial nature of the species listed in Hawaii, but there is far more compelling evidence. As a result of changes made to the Endangered Species Act in 1982, when the FWS develops plans to recover a species, it is required to identify species "that are, or may be, in conflict with construction or other developmental projects or other forms of economic activity."[26] The conflict label is applied *after* a species is listed, but the likelihood of conflict often can be predicted during the listing process either because of comments provided at public hearings or received in response to a proposed listing. In other words, the conflict label serves as a surrogate of prelisting controversy. Development projects and other forms of economic activity occur throughout the United States, so no reason exists to expect anything other than a random geographic distribution of conflict-prone species unless controversy is a factor that affects the willingness to propose that a species be listed. As the data in table 9.1 indicate, however, only noncontroversial species have been listed in Hawaii. About 7.5 percent of all listed species were given the conflict label in 1983, but none of these were in Hawaii. Similarly, the FWS concluded in 1986 that almost 23 percent of the species for which it has responsibility were prone to conflict.[27] Again, however, none were in Hawaii.

Examination of the designation of critical habitats provides still

Table 9.1   The Geographic Location of Conflict-Prone Species, 1983 and 1986 (in percentages)

	Hawaii		All Other States	
	1983	1986	1983	1986
Conflict?				
Yes	0.0	0.0	8.7	26.2
No	100.0	100.0	91.3	73.8
Total	100.0	100.0	100.0	100.0
(*N*)	39	51	242	340

Source: Compiled from data provided by the FWS.
Note: All other states includes Guam, Puerto Rico, the Virgin and Mariana Islands, and the District of Columbia.

another indication that Hawaii's endangered species receive less protection than one might expect based on their numbers and propensity to extinction. As noted in chapter 7, designation of a critical habitat often provokes controversy, although not necessarily for valid reasons. Many people incorrectly believe that the establishment of a critical habitat for a species precludes any use of the habitat for all purposes except for the species' well-being. However justified the opposition, the designation of critical habitats by geographic location offers a second measure of the propensity for opposition. When opposition is a possibility, as in Hawaii, where land is limited and where designation is likely to affect many federal activities, the FWS and the NMFS will be disinclined to propose critical habitats. Once again, a comparison of data from Hawaii and all other states supports this conclusion. Through early 1990, about one-fifth of all native species have had critical habitats designated.[28] In Hawaii, however, only five of fifty-two listed species, or less than 10 percent, have the presumed advantage of a critical habitat. Moreover, for the few Hawaiian habitats designated, the time between the proposal and the actual designation is significantly longer—on average, about four months longer than the time needed to designate habitats in all other states.

Two examples dramatically illustrate the nature of the controversy surrounding critical habitats in Hawaii. The first of these involves the palila, an endangered forest bird. The second involves the Hawaiian monk seal.

The Bird That Went to Court

The palila is found in only one place on earth—the slopes of Mauna Kea on the island of Hawaii. Not only is the bird geographically limited, but it is also completely dependent on naio and mamane trees, without which it cannot survive. The trees provide food, shelter, and nest sites, and the birds' habitat coincides with the location of mature trees. The chief threat to the species' habitat is well known—feral goats and sheep that devour the trees' stems, leaves, sprouts, and seedlings. Eliminating this threat would provide the single most effective remedy for the causes of the palila's endangerment.

The threat to the forests of Mauna Kea has been recognized at least since 1921, when the territorial government initiated a plan to kill as many sheep as possible. Over the next twenty-five years, over 45,000 sheep were killed. Only about five hundred remained in 1950, at which time the territorial government reversed its previous policy and decided to maintain sufficient numbers of feral goats and sheep to provide opportunities for hunters. Due to pressures from these hunters, the state increased the number of feral animals to an "ideal" hunting population of between two and three thousand during the 1960s.[29] Occasionally, when the population dropped below these numbers, hunters complained, and the state managed the feral animals to increase their numbers.

Hawaii's Division of Fish and Game further demonstrated responsiveness to the state's hunters in the 1960s when it introduced mouflon sheep to the Mauna Kea Game Management Area. Mouflons provide excellent meat and horns of trophy quality. The state was aware that the feral animals were destroying the forests of Mauna Kea, but the Division of Fish and Game believed it could regenerate the forest and simultaneously allow the sheep and goats to remain for the hunters' satisfaction.

Hunting is a popular sport in Hawaii and one that the state has long encouraged. The state manages over one million acres—about one-fourth of the entire state—that are available for public hunting. The state prefers that much more land be made available for hunting and that tax incentives be provided to private landowners to induce them to allow hunting on their land.[30] Hunters have access to many non-native species, ranging from boars and deer, to feral goats and sheep, to quails, pheasants, partridge, and wild turkeys. As an indication of hunting's popularity, the number of licensed hunters in Hawaii increased at a much faster rate than did the state's population between 1955 and 1981. The increase in the number of licensed hunters was especially pronounced on the island of Hawaii, where most of the game is located.

The state's efforts to encourage hunting have been successful, but the effects on the palila were devastating. During the 1950s and 1960s, its available habitat continued to shrink, and its numbers were reduced significantly. In response, the palila was one of the first species formally listed as endangered (in 1967). The FWS also identified the palila as one of its ten priority species in 1975, noting that the agency would establish a critical habitat for the species "as rapidly as possible," which it did in mid-1977.[31]

State officials also began to view the situation from a different perspective. Some DLNR employees recommended in 1976 that the feral sheep and goats be eliminated because of the habitat destruction they caused. A year later three divisions of the DLNR, its deputy director, and the chair of the department's governing board made the same recommendation. The governing board refused to act, no doubt due to heavy pressure from hunters, many of whom opposed any measures to benefit the palila.[32] The board said it would authorize the construction of a fence around part of the forest that comprised some of the palila's habitat, but the board did not do so. In fact, the management plan the board developed for the state-owned area in the 1970s perpetuated the hunting.

One might have expected the FWS to persuade the state on the palila's behalf, but the Service was ill-equipped to do so. The FWS realistically appreciated its dependence on the states, especially Hawaii, which has so many endangered species. The Service prefers that the states have primary responsibility for resident species, so the FWS does not want to preempt state control. As Keith Schreiner once explained it, without the help of state conservation agencies, "there is little hope that we will be able to effect the recovery of endangered and threatened species in the United States. . . ."[33] Despite having several options, the FWS apparently decided not to risk its limited political capital in a confrontation with its most important state.

The FWS is also liable for some of the blame for the palila's plight and the continuing destruction of its habitat. For many years the state maintained the mouflon sheep and other feral animals with funds provided through an FWS state grant program, the Federal Aid in Wildlife Restoration Act. Consequently, when the Sierra Club Legal

Defense Fund decided to sue the DLNR on behalf of the palila, the Fund asked the FWS to intervene. The Service declined; its director, Lynn Greenwalt, noted that he preferred such disagreements to be settled out of the courtroom. "I am hopeful," he added, "that the State of Hawaii can be persuaded that the protection of the Palila's habitat is of first importance and that the adverse impacts of feral goats and sheep on Mauna Kea must be eliminated." [34]

In its lawsuit in federal district court, the Sierra Club Legal Defense Fund (and several coplaintiffs, which included the palila) claimed that the state's actions constituted an illegal "taking" of the palila since the feral animals were destroying the birds' only known habitat. The state rejected this argument, asserting that the number of birds had increased, that these numbers could increase still further, and that no evidence existed that the palila depended solely on the naio and mamane trees inasmuch as captive breeding had never been tried. The court rejected each of these assertions and noted the state's "demonstrated susceptibility" to hunters. As a result of its conclusion in *Palila* v. *Hawaii Department of Land and Natural Resources* [35] that the feral goats and sheep were having a devastating effect on the forest, the court ordered their complete elimination within two years of its decision in June 1979.

Rather than accept the verdict and act to protect the palila, the state appealed the district court's decision. A federal appellate court rejected the appeal in early 1981,[36] but five years later more than 150 feral sheep still roamed the palila's critical habitat. When additional research showed that the mouflon sheep were also destroying the mamane trees, the Sierra Club Legal Defense Fund asked that these sheep too be removed. The state refused, and so the Defense Fund and its coplaintiffs sued a second time on behalf of the palila. The species was again victorious, despite the state's belief that, through careful management and oversight, it would be "possible both to maintain a viable sport-hunting population of mouflon and to enhance the mamane ecosystem to encourage survival of the Palila." [37] In order to enhance its prospects for survival, however, the district court ruled in late 1986 that the mouflon too would have to be eliminated from the critical habitat of the palila.

Accommodation and compromise are not limited to the palila. Indeed, Hawaii's experience with endangered species provides another example of the difficulties in achieving agreement among governmental organizations that do not share similar goals.

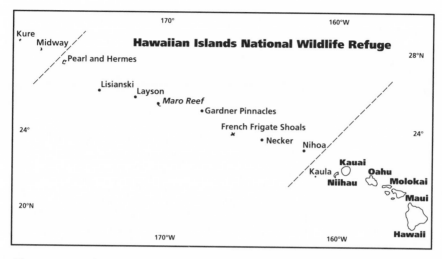

Figure 9.1   The Hawaiian Islands. Source: FWS Region 1, *Hawaiian Islands National Wildlife Refuge: Master Plan/Environmental Impact Statement* (Honolulu, 1986): 0.1.

The Seal That Came Up Short[38]

Only three species of monk seals are known to humans. One, the Caribbean monk seal, is believed to be extinct. The last confirmed sighting occurred in 1952. The Mediterranean monk seal still survives, but is found only occasionally and then in small numbers on isolated beaches in Europe and North Africa. Given its few survivors and the level of pollution and human activity in the Mediterranean, the outlook for these seals is not favorable. The third species, the Hawaiian monk seal, is now found only on the small and remote leeward Hawaiian islands, northwest of Kauai (see figure 9.1). These islands and atolls, from Nihoa to Kure Atoll, extend over eleven hundred miles, but their total area is less than seven square miles.

The seals' habitat provides an important clue to their well-being. The Hawaiian monk seals are literally living fossils. Scientists estimate that the seals' isolation has left them virtually unchanged over the last fifteen million years.[39] Largely free from intrusions of all kinds, the seals have not developed defense mechanisms. Unfortunately for the seals, this complacency threatens their survival. The seals do not flee from humans, so they are easily killed, as thousands of them were in the 1800s and early 1900s. More important, the seals are extraordinarily sensitive to any kind of human intrusion. One seal expert has emphasized that "the mere presence of human intruders among the 'tame' seals appears to cause mortality among newborn pups." Others have concluded that the seals "probably cannot coexist with man."[40] The available evidence supports these observations. Although the seals once inhabited the major Hawaiian Islands, such as Oahu and Kauai, seals are rarely sighted there. In the northwestern Hawaiian Islands scientists noted a considerable reduction in the number of seals between 1960 and 1980. The largest reductions came on those islands with the greatest frequency of human intrusions. Precise numbers are not available, but less than one thousand monk seals probably remain, making the Hawaiian monk seal one of the most endangered of all marine species.

Due to the precipitous decline in the number of seals and the increased human activity in their habitat, a government advisory body, the Marine Mammal Commission (MMC), recommended to the NMFS in 1975 that the species be listed as endangered.[41] In the commission's view, however, merely listing the species would not be enough. A critical habitat should be designated for the seals because "protection of habitat, including the island beaches and surrounding near shore water, from all forms of human intrusion may represent the single most important action that can be initiated to protect the species from extinction."

However straightforward this recommendation, it inaugurated a series of bitter debates that continued for more than ten years. The major contestants included the FWS, the NMFS, the MMC, and the state of Hawaii. Each has interests that frequently conflict. The FWS manages the Hawaiian Islands National Wildlife Refuge, which President Theodore Roosevelt created in 1909 to protect seabirds from being slaughtered for their feathers. The refuge includes all the islands from Nihoa to Pearl and Hermes Reef (see figure 9.1). The seals use the islands in the refuge for resting and pupping and their offshore waters for feeding and breeding. The FWS has an interest in protecting the seals, but it also has a statutory obligation to allow some human use of the refuge in order to provide some benefits to the public.

The NMFS has a statutory mandate to encourage the development of commercial fisheries, but it also has the job of protecting the seals because they are marine mammals. Congress created the MMC in 1972 as an independent agency to review federal activities involving marine mammals, including NMFS implementation of the Marine Mammal Protection Act.[42] The MMC has no regulatory authority and normally must rely on its powers of persuasion and the weight of the evidence it can marshal to sustain its recommendations to other federal agencies.

The last governmental actor, the state of Hawaii, has several interests. At the least, the state disagrees with the FWS definition of the boundaries of the Hawaiian Islands National Wildlife Refuge. The state insists that it owns about 250,000 acres of submerged lands that the FWS includes in the refuge.[43] The disputed jurisdiction means that Hawaii is especially sensitive to what it considers to be federal preemption of state responsibility. The state government also intends to develop fisheries as a way of diversifying its economy. In 1979, for example, the state published a plan designed to stimulate commercial fishing.[44] This plan and a revised one issued in 1985 identify the northwestern Hawaiian Islands as prime fishing locales. To accommodate the expected increase in commercial fishing, onshore support facilities would be necessary, but the FWS has opposed them, and commercial fishing remains prohibited within the refuge.

These organizational characteristics are important in understanding the fate of the monk seal. There was no disagreement among the various agencies that the seal should be listed as endangered, and this was accomplished in late 1976, when the FWS and the NMFS jointly listed the species.[45] The contentious issue focused on the need for an appropriate size of a critical habitat. Disruption of habitat is the major threat the species faces. The MMC had recommended that a critical habitat be designated and, prior to the listing of the seal, so also had the FWS regional office with responsibility for Hawaii. This regional office emphasized that "uncontrolled commercial fisheries exploitation" created serious potential problems.[46] In spite of these recommendations, when the seal was listed, the NMFS indicated that no critical habitat would be designated at that time because no one had requested it![47]

Two weeks after this decision, the MMC again recommended to the NMFS the designation of a critical habitat. In the commission's view, an ideal critical habitat would include all lagoons, beach areas, and seaward waters up to three miles from six of the northwestern islands. Once the habitat was designated, the MMC emphasized that all human intrusions, including fishing, should be terminated "to the greatest extent possible." After nearly a year's delay, the NMFS conceded that it was essential to establish a critical habitat, but suggested that, in addition to the beaches and lagoons, its outer limit include ocean waters but only to a depth of ten fathoms. This would be considerably less area than the MMC had recommended, but the NMFS accepted the need to place restrictions on the use of fishing gear well beyond the ten-fathom boundary.

Although the NMFS had decided that a critical habitat was imperative, it took no action to propose one. A third time, in early 1978, the MMC again advocated a three-mile limit, but its recommendation had no apparent effect. With increasing frustration and in a belief that the seals were in "grave danger of extinction," the MMC complained that its recommendations were being ignored and that NMFS efforts on behalf of the monk seal were inadequate.[48]

Some movement did occur in February 1980, when the NMFS issued a draft environmental impact statement, which discussed the possible consequences of three alternative critical habitats, including outer limits of ten and twenty fathoms (covering 367 and 2,041 square nautical miles, respectively) and three miles (covering 794 square nautical miles). The NMFS, while acknowledging that designation of a critical habitat was essential, did not express a preference for any of the three options. It pointed out, however, that a ten-fathom boundary would be insufficient "for the health, well-being and continued viability" of the monk seal.[49] The Honolulu office of the NMFS expressed its preferences when it advocated a twenty-fathom boundary.

After several years of discussion and a specific recommendation from NMFS officials in Hawaii, one might have expected that a critical habitat would soon be proposed. Such was not the case, and the NMFS continued to defer a proposed designation. Arguing that its experts could not agree on the proper boundaries for a critical habitat, the NMFS said it needed more data before acting. These experts returned their opinion in 1983. They urgently recommended the designation of a critical habitat and argued that it should extend to the twenty-fathom line.

At this point the NMFS office in Honolulu prepared a draft "decision memorandum" that included the twenty-fathom boundary. The draft justified the boundary and indicated that it would encompass "those land and water areas considered essential to the survival as well as recovery of the Hawaiian monk seal."[50] The agency's assistant administrator for fisheries concurred with the recommendation in May 1984. After nearly nine years of discussion and discord, a decision had finally been made to propose a critical habitat.

If it is possible to believe, action was further delayed by the NMFS's release of a supplemental environmental statement, which listed the ten-fathom line as the preferred alternative. Eventually, however, the NMFS proposed a critical habitat with a boundary of ten fathoms in early 1985. The FWS and the MMC protested the choice and argued for twenty fathoms. The NMFS's monk seal experts also rejected as inadequate the ten-fathom boundary, complaining that it was inconsistent with biological data and indefensible in terms of the seals' feeding habits and ecological requirements. Even the head of the NMFS office in Honolulu objected to the choice and several times brought his views to the attention of his superiors. In contrast, the governor and the DLNR remained unconvinced of the need for the critical habitat.[51] The data, the state emphasized, simply did not justify any further protective measures for the seal.

After waiting nearly a year for the final rule to be published, the Sierra Club Legal Defense Fund and Greenpeace International filed suit against the NMFS in early 1986. Two months later, the NMFS released a final environmental impact statement that discussed the three different boundary options for the proposed designation of the critical habitat.[52] The statement concluded that the twenty-fathom boundary was not likely to have any adverse economic impact on state or private activities and that the environmental impacts would be the same as the impacts associated with the ten-fathom line. Ironically, on the same day that the statement was released, the NMFS also published the final rule on the critical habitat, which established a ten-fathom boundary.[53]

The MMC again protested the choice, labeling it as biologically inappropriate. The commission asked the NMFS to reconsider its decision, but it was completely unresponsive. The Sierra Club Legal Defense Fund and Greenpeace amended their earlier suit and claimed that the ten-fathom designation was inadequate. Both the plaintiffs and the NMFS subsequently filed motions for summary judgment, both of which were rejected. The NMFS had staved off defeat, but not criticism. At a hearing on the motions, the judge revealed his distaste for the NMFS's position and intransigence.[54] Faced with an unsympathetic judge and the likelihood that its own scientists would testify that the ten-fathom boundary was scientifically indefensible, the NMFS eventually capitulated. It extended the boundary of the critical habitat to twenty fathoms, notwithstanding objections from the state of Hawaii.[55]

Lessons

This assessment of interorganizational implementation is instructive in at least several ways. First, the illustrations demonstrate that administrative promptness is not a virtue associated with interorganizational relations. Once the MMC recommended the designation of a critical habitat for the monk seal, for example, more than ten years elapsed before action was taken. Even then, the initial action was inadequate. In the end, the threat of legal action inspired the NMFS to act sooner than it had intended or preferred. The nature of the threat to the palila has been known for decades, yet the threat still exists more than twenty years after the species' listing as endangered.

In each case, knowing the likely solution but allowing the threat to continue unabated demonstrated the political nature of the decision-making process. The technology to solve the species' plight is largely available, but the political will to do so is problematic at best. Viewed from another perspective, the situation with the species discussed in this chapter may not be perceived as problems at all. Organizations selectively choose the lens they wear in perceiving items on their institutional agenda. How an organization views a condition depends on its goals, perspectives, and the political environment in which it operates. Conditions, such as endangerment, do not become problems until people or organizations define them as such. "Conditions become defined as problems," John W. Kingdon explains, "when we come to believe that we should do something about them."[56] Conditions are tolerable situations whereas problems are irritants that people seek to remedy. Quite clearly, then, the NMFS view of the sea turtles and the monk seal and the state view of the palila illustrate perception of a condition rather than recognition of a problem. In this regard, the story of these species is probably typical of many other endangered species throughout the United States.

Second, the species discussed in this chapter exhibit the conflicting pressures that many administrative agencies face. However desirable it is to protect endangered species, their fate is rarely independent of the agencies that have an interest in them. As an illustration, Hawaii's Endangered Species Act mandates that all federally listed species in the state will be considered as endangered under state law. All such species are supposed to be the beneficiaries of "positive actions to enhance their prospects for survival."[57] At the same time, however, Hawaii's DLNR has a mixed, but not necessarily overlapping constituency. It includes hunters, conservationists, commercial fishermen, and land developers, to list only a few. A natural inclination among bureaucratic organizations is to balance competing demands. The DLNR tried to do this in serving the interests of the palila and of the state's hunters. The state policy of coexistence attempted to maintain, simultaneously, a huntable population of feral animals and a suitable habitat for the palila.

In addition, while attempting to balance competing demands, government agencies can ill afford to neglect powerful and well-organized constituents. Throughout the years of debate over a critical habitat for the monk seal, the DLNR faced intense pressures from commercial fishing interests. So effective has this pressure been that at least one close observer of the situation, Craig S. Harrison, has stressed that the "DLNR is so strongly committed to fishing development that it has become a *de facto* single-purpose agency on this issue."[58] It was not unexpected, therefore, that the state complained to the MMC that the establishment of a critical habitat for the seals would lead to "the possible loss of a potentially large fishing industry. . . ."[59] One might have expected that the state's major conservation agency would give the seals the benefit of the doubt, but this was not the case; for many years the state strongly opposed the designation of a critical habitat for the monk seal. And, as an examination of the record shows, the NMFS's response to this opposition was accommodation in the form of delay.[60]

Accommodation is an option when compromise is a possibility and when the disputants can be satisfied with less than complete victory. Both accommodation and compromise are at the heart of political decisionmaking. At least one Commerce staff member acknowledged the monk seal decision as just that. Before the NMFS issued the final environmental impact statement on the monk seals' habitat, one official in the National Oceanic and Atmospheric Administration, to which NMFS is subordinate, characterized the debate in this way: ". . . the Governor and the Fishery Management Council don't want any critical habitat designation; the Marine Mammal Commission, the [monk seal] Recovery Team, and environmentalists want critical habitat designated out to the 20 fathom line. The 10 fathom line is simply a compromise."[61] Unfortunately for many endangered species, their needs recommend against such compromise because compromise can spell demise.

Third, relative passivity best characterizes the NMFS in several of the interorganizational debates. Why, for example, after deciding that the sea turtles were endangered, was the FWS willing to negotiate with the NMFS about which agency would have responsibility for the species?[62] Once these negotiations started, why were they so protracted, especially since both agencies had already agreed to list the turtles? While appearing to be genuinely concerned about the palila, the Service has also demonstrated a sensitivity to Hawaii's DLNR. There is evidence that the FWS attempted to persuade Hawaii to be more responsive to the needs of the palila, but the FWS seemingly tailored some of its actions to mollify the state. For example, the FWS appointed a team of experts to make recommendations about what should be done to protect the palila. The team said that an ideal critical habitat would encompass the bird's entire historical range.[63] In deference to the state and its hunters, however, a much smaller habitat was designated.

Similarly, in none of the legal actions involving the palila did the FWS intervene. In fact, after the initial judicial decision in 1979, in which the court ruled that a "taking" of the palila had occurred through harm to its habitat, Interior proposed a redefinition of harm. The existing definition, and the one on which the 1979 decision was based, was broad and defined harm in such a way as to include acts that significantly disrupt an endangered species' life-style or that significantly degrades its environment.[64] The proposed change, which was specifically intended to preclude future decisions similar to the one reached in the palila case, would have restricted the meaning of harm to require death or injury to individual members of a listed species. So unpopular was the proposal that Interior altered the original and provided a definition only marginally different from the existing one.[65]

Last, and perhaps most obvious, as the number of institutional actors increased, so also did the number of contested issues and the level of disagreement about what should be done. With the monk seal, the FWS, the NMFS, and the MMC were in agreement that the scientific data justified a critical habitat and that existing levels of protection were inadequate. The three organizations disagreed about how large the critical habitat should be and when it should be designated. The DLNR took an entirely different position on these issues. It claimed that an insufficient biological basis existed to justify designation. Moreover, the DLNR argued that existing state regulations already provided suitable protection for the seals. Similar disagreements about the quality of scientific data have also characterized several efforts to list plant species in the state. In a few instances as well, issues completely unrelated to endangered species have colored reactions. A consistent complaint from the state has been that federal actions preempt Hawaii's right to manage its lands and waters.

The issue of preemption is an important one because it places the FWS or the NMFS in the position of directly affecting, or at least attempting to influence, the actions of other organizations. For the FWS and the NMFS, the listing of species and the designation of critical habitats is a statutory obligation, yet the exercise of this obligation is often tempered by other considerations. How these considerations affect other federal agencies' actions in regard to listed species and critical habitats is the subject of the next chapter.

All of the effort associated with the implementation of the Endangered Species Act is ultimately directed at a single goal —the recovery of endangered species to the point where their continued existence is no longer in doubt. Listing, designation of critical habitats, and consultation are only interim steps. By way of analogy, listing a species as endangered is comparable to admitting a patient to a hospital's intensive care unit. Admittance is an indication that the patient is in critical condition, just as listing signifies the same for an endangered species. Designation of a critical habitat and consultation may be helpful, but they address only some concerns, namely those that result from federal activities or federally licensed or permitted projects. Moreover, these two processes are intended to mitigate rather than to eliminate the problems facing listed species. What is needed is a cure so that species in jeopardy can recover and be delisted. Occasionally all that a listed species needs to recover is time to regenerate or to populate a new or formerly insecure habitat. More likely, a cure will involve the application of remedial measures. Such measures can include research, transplantation, land acquisition, public education, captive or artificial breeding, special management practices, or whatever botanists and wildlife biologists deem necessary to bring about a species' recovery.

Unlike human patients, who are normally assured of receiving appropriate attention, many endangered species are less fortunate. Just as limited staff and resources adversely affect other protection efforts, these same constraints limit recovery operations. Quite simply, the costs of recovering listed species far exceeds the federal government's willingness to commit the necessary resources or to impose constraints on activities that jeopardize listed species. Consequently, some choices must be made. Not all listed species can be the beneficiaries of remedial measures or even of a proper determination of what these measures should be. Once again, the issue is one of priorities. Which species will be allowed to languish in intensive care and perhaps disappear because of neglect? Which species can recover with only limited attention and assistance? Which species should capture attention and resources, and thus find their prospects improved? The brief discussion that follows provides some insight into how these questions can be answered. The discussion parallels the earlier discussion of priorities found in chapter 4, but here the emphasis is on recovery efforts.

At least four criteria are available to assist in answering the questions above. The first two criteria, degree of threat and potential for recovery, are based on biological judgments. Some listed species face a greater threat to their existence than do others. Some wildlife biologists believe, therefore, that the degree of threat should determine priorities for recovery.[1] Making choices on the basis of degree of threat means that those species most in need of attention or intervention would receive it first. This makes intuitive sense until one realizes that the probability of a successful recovery effort is another relevant variable. Ideally, species facing a high threat would also have a high potential for recovery. Some scientists assert, however, that the most highly endangered species are the least likely to benefit from recovery efforts because these are the species closest to extinction.[2] For such species the chances of success are low, and there is a high probability that scarce resources will be wasted should the recovery effort fail. As seen in figure 10.1, having only two levels of threat and two levels of recovery potential creates several choices.

Species found in the upper left quadrant would be priority candidates for recovery efforts while those in the lower right quadrant would be unlikely to receive any immediate attention. In contrast, species found in the remaining quadrants (i.e., low threat–high recovery potential and high threat–low recovery potential) would present more difficult choices. These choices are compounded when the number of levels is increased, perhaps by having high, medium, and low threats. More significant, one must realize that judgments about threat and recovery potential are not objective scientific determinations.

Degree of Threat

High   Low

Recovery
potential

High

Low

	High	Low
High	1	?
Low	?	4

Figure 10.1   Relating Threat and Recovery Potential

A problem in relying solely on threat and recovery potential is that
they both neglect considerations of cost, an important concern when
resources are limited. Is it worthwhile to spend $2 million trying to
recover one high-threat species with a high recovery potential when
the same resources could be used to assist in the recovery of twenty
low-threat species with high recovery potential? Or, as an economist
concerned with cost-effectiveness might suggest, money should first
be allocated to those species with the lowest costs of preservation.
Following this dictum, species with the lowest recovery costs would receive initial attention. Resources would be devoted to species with the
highest estimated costs only after all less expensive species received
attention, thus maximizing the number of species that could be saved
for a given level of resources.

A related perspective would attempt to ensure that the benefits of a
species' recovery would exceed the costs of recovery. The consequence
of this choice is that species with few benefits, however measured,
would not be recovered if they also entailed high recovery costs.
Though both of these perspectives are economically rational, they
exclude the scientifically relevant consideration of degree of threat.

The last set of criteria or values includes whatever else remains.
That is, endangered species may find their path to recovery blessed
because of their political, cultural, symbolic, or ecological significance.
The bald eagle provides a good example. An astute administrator
may recognize that he cannot easily disregard the needs of the eagle,
even when this species may not be in dire peril. Similarly, this administrator might assert the scientific and ecological merits of recovering
monotypic genera and keystone species before subspecies, even when
the latter faces a high threat. Some combination of all the criteria

is possible, but one can readily appreciate how cumbersome such a system would be.

Initial Perspectives

Efforts to recover selected endangered species predate the passage of the Endangered Species Preservation Act of 1966. In the early 1940s, as an illustration, scientists estimated that less than twenty whooping cranes survived, from a population that once numbered about fifteen hundred. After a search of several years' duration, the cranes' nesting grounds were located in northern Canada in 1954, thus allowing the beginning of a program to increase the cranes' numbers to the point where they now exceed 140 individuals.

Despite the early attention devoted to the recovery of whooping cranes and a few other endangered species, recovery programs did not represent a major thrust of Interior's activities, at least through the mid-1970s. The first draft recovery plan was not available until 1972, and nearly three more years would pass before final approval of the first plan, in April 1975, for the California condor. According to the FWS, a recovery plan is intended as a

> guide that delineates, justifies, and schedules those actions re-
> quired for restoring and securing an [endangered or threatened]
> species as a viable self-sustaining member of its ecosystem. The
> prime objective of all recovery plans is to provide a scheme that,
> if implemented, will lead to the improvement of the status of the
> species to the point where the species qualifies for delisting. . . .
> A typical plan should be built around a step-by-step outline of
> actions necessary to reduce or resolve the problems or limiting
> factors which contribute to the listed status of the species.[3]

The plan for the condor was expected to be the first of many. Shortly after it was published the FWS declared its intention to have recovery plans finished for scores of species within just a few years. This quickly proved to be no more than an overly optimistic projection. Through the end of 1978, only seventeen recovery plans covering twenty species were finished. What is noticeable in reviewing these plans is that the FWS did not base its efforts on a readily identifiable priority system.[4] Birds and fishes were the subject of three-quarters of the initial plans. These plans focused on species facing both high and low threats, a species believed to be extinct (the blue pike), another species, the condor, considered well on its way to extinction, and on species thought to have excellent prospects for recovery. The FWS had focused on only twenty species, so many other species with a high threat *and* a high recovery potential were neglected altogether. This neglect was probably not intentional. Development of recovery plans requires highly specialized expertise and a thorough familiarity with a listed species' needs, habitat, life-style, population parameters, and trends. Much of this expertise is not located within the FWS or the NMFS, but rather in museums, universities, and state wildlife agencies. This meant that the FWS and NMFS often had to rely on teams of volunteers to prepare recovery plans. This approach had its advantages, but it likewise meant that the FWS in particular was dependent on the team members' willingness to coordinate their work schedules with other volunteers and on their sense of professional obligation to finish their tasks. Reliance on this sense was not always successful —many recovery plans for which volunteers had responsibility were never finalized or were many years late. A recovery team for the San Joaquin kit fox was appointed in April 1975, but a plan for the species was not approved until January 1983.

The delay in approving plans was not, however, entirely the fault of unreliable or delinquent volunteers. The FWS received a "technical draft" of a plan for whooping cranes in March 1976, but the plan did not receive final approval from the agency's director until January 1983. Delays of three to six years also prevented prompt approval of technical drafts for other species.[5] Plans for several of these species were supposedly receiving priority attention.

Until 1978 the FWS could also defend its record on recovery plans by noting that no such plans were legally required, so it devoted few resources to them. This changed with the passage of the Endangered Species Act Amendments of 1978, which mandated the development and implementation of a recovery plan for every listed species unless such a plan would "not promote the conservation of the species." The following year Congress required the development of a priority system for recovery efforts.[6] The FWS had been developing such a system, so the 1979 legislative requirement did not impose much additional work. In 1980 the FWS adopted a recovery priority system that gave preference to species and subspecies subject to a high degree of threat with a high recovery potential, followed by high-threat, low-recovery potential species and subspecies. The legislative amendments approved in 1979 also apply to the NMFS, but it ignored the statutory mandate for ten years before publishing a priority system for the development of recovery plans.

The ground rules for the development of these plans changed once again with the passage of the Endangered Species Act Amendments of 1982. The amendments require the FWS and the NMFS to give priority in the preparation of recovery plans, to the "maximum extent prudent," to species "that are, or may be, in conflict with construction or other developmental projects or other forms of economic activity."[7] Both the Interior and Commerce Departments opposed the provision when it was under congressional consideration. The departments declared that the requirement could work to the detriment of some species most in need. As a Commerce spokesman explained, the conflict provision might "require that priority be given to species with stable or increasing populations rather than species with seriously declining populations simply because the more healthy species is or may be in conflict with certain projects."[8] Congress ignored this concern, which meant that a revised priority system would have to be developed, published in the *Federal Register*, subjected to public comment, and then issued in final form.

In reality this meant that the FWS would be adopting its third recovery priority system in three years. Shortly after the beginning of the Reagan administration the FWS had revised its existing system, which it had formally adopted only a few months earlier. The new, or second system, favored the recovery of higher life forms, namely mammals, birds, and fishes over plants, insects, and other invertebrates. According to this plan, for example, the recovery priority for mammalian subspecies with a low recovery potential would be much higher than for insect and amphibian species with high recovery potentials. This approach did not long survive, and the FWS recovery system proposed and eventually adopted in 1983 made no distinction between higher or lower life forms.

The system, which is displayed in table 10.1, closely parallels the original FWS recovery priority system. The degree of threat and recovery potential are the two most important criteria, with preference then given to monotypic genera over species and subspecies.[9] Neither of the two earlier systems had given any preference to monotypic genera, but the FWS believes that such a preference is warranted in order to preserve maximum genetic diversity. Conflict is an additional variable that must now be considered. Species are given a conflict des-

Table 10.1 FWS Recovery Priority Guidelines

Degree of Threat	Recovery Potential	Taxonomy	Priority	Conflict
High	High	Monotypic genus	1	1c / 1
High	High	Species	2	2c / 2
High	High	Subspecies	3	3c / 3
High	Low	Monotypic genus	4	4c / 4
High	Low	Species	5	5c / 5
High	Low	Subspecies	6	6c / 6
Moderate	High	Monotypic genus	7	7c / 7
Moderate	High	Species	8	8c / 8
Moderate	High	Subspecies	9	9c / 9
Moderate	Low	Monotypic genus	10	10c / 10
Moderate	Low	Species	11	11c / 11
Moderate	Low	Subspecies	12	12c / 12
Low	High	Monotypic genus	13	13c / 13
Low	High	Species	14	14c / 14
Low	High	Subspecies	15	15c / 15
Low	Low	Monotypic genus	16	16c / 16
Low	Low	Species	17	17c / 17
Low	Low	Subspecies	18	18c / 18

Source: *FR* 48 (21 Sept. 1983): 43104, as corrected in *FR* 48 (15 Nov. 1983): 51985.

ignation whenever a federal agency identifies a listed species that is or may be in conflict with one of its projects or when a consultation under Section 7 produces a negative biological opinion or a recommendation of prudent and reasonable alternatives to avoid a negative biological opinion.

When the FWS first proposed the current system in April 1983, it generated some interesting reactions, especially from those within the FWS. The Service's goal is to improve the status of endangered or threatened species so that they can be delisted.[10] To some people within the FWS this suggested that attention should be devoted to species facing a low threat and, therefore, those close to recovery. In fact, however, the system lowers the priority for the species closest to recovery. "The greatest problem with the recovery implementation system," complained one FWS employee, is "that close adherence to it will assure we seldom recover and delist a species."[11] The employee similarly objected to the system because he believes it does not allow a cost-effective use of limited resources. These resources, he argued, are used most effectively when recovery plans are directed at species with high recovery potential before any species with a low recovery potential. This objection has at least some merit in view of the difficulties associated with the recovery of the latter species. The threats to their existence are either poorly understood or "pervasive and difficult to alleviate."[12] The result is a need for intensive management, reliance on experimental or untried techniques and, most important, a highly uncertain outcome. There is, however, an alternative perspective, to which the FWS agreed. This perspective argues that recovery is less important than prevention of extinction.[13] According to this view, resources should be directed toward preventing extinction rather than be used to assist highly recoverable endangered species confronting a low threat. And, of course, this approach also decreases the likelihood that endangered or threatened species will be recovered.

Some FWS employees also believe that nonscientific considerations should be incorporated into decisions about which species should have recovery plans and resources to implement them. For many years the FWS has taken the position that political or socioeconomic variables should not influence the development or implementation of recovery plans. Some FWS officials disagree. They argue that species with high visibility, national significance, and sociopolitical value should be given some preference, supposedly to generate public and political support for the agency's efforts.[14] This contention again provides explicit recognition that some species are believed to be more important politically than others, at least in the eyes of those responsible for ensuring the continued existence of listed species. From a bureaucratic perspective, such a bias is not altogether undesirable. This is particularly true for an agency like the FWS that must justify its spending preferences not only to political appointees in Interior, but also to the budget-conscious OMB and appropriations committees in Congress. In brief, there are biological justifications (e.g., degree of threat, recovery potential, and taxonomy) that direct attention to certain species while other, nonbiological values (e.g., political, cultural, and symbolic ones) encourage attention elsewhere.

Assessing the Development of Recovery Plans

There are conflicting or overlapping pressures in regard to the development of recovery plans, but the data are available to determine which values predominate. If the FWS favors biological criteria then the development of recovery plans should parallel the Service's recovery priority system (see table 10.1). Thus, the following propositions can be tested:

1. Class (i.e., mammals, birds, fishes, and so on) is an irrelevant variable, and no discernible pattern that favors one class over another in the development of plans should be evident.
2. Species facing high threats should have plans prepared before species facing low threats. Degree of threat is supposedly related to listing as either endangered or threatened, so the former should get plans prior to threatened species.
3. Species with a high recovery potential should have plans prepared for them before species with low recovery potential.
4. Full species should have plans prepared prior to plans for subspecies.
5. Conflict-prone species should have recovery plans developed before species that do not face conflicts with developmental activities.

Sharp divergences from these projections will provide evidence suggesting that nonbiological variables play a role in the development of recovery plans.

Ideally, a single, comprehensive table would display the data addressing the five propositions. However desirable, such a table would

Table 10.2 Listed Species with Recovery Plans

	Mammals	Birds	Reptiles & Amphib- ians	Fishes	Snails, Clams & Crusta- ceans	Insects	Plants
With Recovery Plans (%)	43.6	69.5	71.8	59.7	68.8	60.0	45.0
Without Recovery Plans (%)	56.4	30.5	28.2	40.3	31.2	40.0	55.0
Total (%)	100.0	100.0	100.0	100.0	100.0	100.0	100.0
(N)[a]	55	82	39	77	48	20	198
Plans Approved in:							
1975–77	2	3	1	3	0	0	0
1978–80	5	12	2	9	0	1	2
1981–83	10	24	6	12	7	2	13
1984–86	4	16	18	18	22	9	39
1987–89	3	2	1	4	4	0	35
Totals	24	57	28	46	33	12	89

Source: Compiled from data provided by the FWS.

[a]The totals include all species listed as endangered or threatened on 31 December 1988, and all recovery plans approved through 31 December 1989. Some species are listed as both endangered and threatened depending on their location. Such species are counted as a single species although they may have more than one recovery plan.

disguise more information than it might disclose because both the degree of threat confronting a species as well as its recovery potential are subject to change. Data on these two variables are available for three years—1978, 1983, and 1986. Moreover, two agencies, the FWS and the NMFS, have responsibility for preparing recovery plans, some plans take years to prepare, and a sufficient number of exceptions exist so that reliance on a single table might be misleading. Accordingly, separate analyses are both necessary and desirable.

The data in table 10.2 reveal that, for all species listed through 1988, recovery plans had been prepared for 56 percent by the end of 1989. Most of these plans were finalized between 1982 and 1985, a period in which the FWS shifted emphasis to the development of plans at the expense of listing additional species. This emphasis proved to be short-lived. After having approved plans for seventy-nine species in 1984, the FWS approved plans for only sixty-four over the next three years.

To put the number of approved recovery plans into some perspective, it is useful to consider projections for the number of plans the FWS intended to develop. These projections have been quite ambitious. When he was the director of the FWS, Lynn Greenwalt once indicated that an "efficacious recovery program" would require the formulation of three hundred plans between 1 October 1978 and 30 September 1980.[15] Only twenty-three plans were completed during these two years notwithstanding Greenwalt's optimistic projections.

Part of the problem in achieving the desired number of plans involves the time required to prepare them once a species is listed. The development of a recovery plan usually does not begin immediately after listing, but the date of listing does provide a convenient benchmark. It is at this point that the FWS or the NMFS formally declares that a species is in jeopardy and that it needs remedial assistance. For the typical species with a recovery plan, the median interval between its listing and the approval of a plan exceeds six years (see appendix table C.1). For a few species the interval is as much as seventeen years; few species, only about 25 percent, are accorded approved recovery plans within four years of their listing. The long delays mean that as late as 1985 the FWS was still finishing recovery plans for species listed in 1967, when Lyndon Johnson was president. In fact, nearly 60 percent of the plans approved during the Reagan administration were for species listed *prior* to the passage of the 1978 Endangered Species Act Amendments (see appendix table C.2). This pattern is likely to continue. A few species listed in 1967 and many in the 1970s did not have recovery plans approved at the end of the 1980s.

With passage of the Endangered Species Act Amendments of 1988, Congress also assured that many of these species would not have plans approved until the 1990s. Following an increasingly typical pattern, the amendments limited the Service's bureaucratic discretion in developing recovery plans. Procedural requirements in the law require the secretaries of commerce and the interior to provide public notice and an opportunity for public review and comment before they approve any new or revised recovery plans.[16] Before this approval is given, all information provided during the comment period must be considered not only by the relevant secretaries but also by each federal agency potentially involved with the recovery effort. The consequence of this procedural change was not difficult to detect. Twenty-six plans were approved in 1988, but only six in the year following passage of the new amendments.

In terms of the first proposition, the data in table 10.2 also indicate the distribution of recovery plans among different taxa. Birds, clams, snails, reptiles, amphibians, and crustaceans are favored species. Initial appearances notwithstanding, some caveats are in order. The data include native species for which the FWS and the NMFS share responsibility (six sea turtles) as well as for species for which the two agencies have sole responsibility. Among U.S. species, the NMFS has sole management responsibility for eight endangered whales, three seals, and two fishes. For only one of the thirteen species, the Hawaiian monk seal, had a recovery plan been developed by 1989. The NMFS has been particularly recalcitrant in its efforts, at least when one considers that it agreed to prepare a plan for humpback whales only after the Sierra Club Legal Defense Fund threatened to sue the agency for failure to do so.[17] As early as 1978 the humpback whale had been identified as a species with a high threat and a high recovery potential, thus being a prime candidate for a recovery plan. A draft plan was finally issued in late 1989.

If one excludes the seals and whales for which the NMFS has responsibility, the percentage of mammals with recovery plans is just about the same as for all other species. This leaves only one category, plants, in which the percentage with recovery plans is well below the average for other taxa. Here the explanation lies in the fact that dozens of plants were listed in the mid- and late 1980s, so it would be premature to expect that plans for many of these species would be finished so soon thereafter.

In short, the distribution of recovery plans among different taxa does not favor one over another. Nonetheless, further examination of the data reveals a somewhat unexpected pattern. During the Carter administration, higher rather than lower life forms were far more likely to have recovery plans developed. In contrast, a more equitable distribution of effort is evident subsequent to 1980. The Reagan administration initially expressed a preference for higher life forms in assigning priorities to listing and recovery efforts, but this preference is not readily evident in regard to recovery plans. Lower life forms like snails, plants, insects, and crustaceans fared well in the mid-1980s. Reagan's Interior Department finalized the first recovery plans for snails, clams, and crustaceans, over 90 percent of the plans for insects, and sixty-three of the first sixty-five plans for plants.

Last, the number of species with recovery plans identified in the table differs slightly from the number of species that the FWS reports. The FWS lists the eastern timber wolf as a single subspecies, but the Service has three separate plans for three different subspecies of wolves. For purposes of determining how many species have recovery plans, the FWS counts the three plans as if they applied to three separate species, rather than one species as reflected in table 10.2. As strange as it may seem, the monthly tally of recovery plans, which is included in each issue of the *Endangered Species Technical Bulletin*, long counted a plan for the blue pike, a species the Service declared to be extinct in 1983. Such a plan does exist, but the justification for counting a plan that was never implemented is questionable.

The second and third propositions focus on degree of threat and recovery potential as priority criteria. Briefly, species facing the highest threat and having the best prospects for recovery should have plans developed before species facing moderate or low threats or poor prospects for recovery. If this is an accurate reflection of administrative efforts, then the percentages of species with recovery plans should decrease as one moves from high-threat/high-recovery potential species to low-threat/low-recovery potential species. The data in table 10.3 examine whether this is the case by collapsing the threat and recovery criteria (from table 10.1) into six categories. The ordering of each of the columns in the table parallels the FWS recovery priority system for the three years in which data are available. For each of the three years the percentages found in the first column of figures should be higher than the percentages in all adjacent columns. This is not the case, and it means that many low priority species get plans before those in greater need.

For some species the threat that confronts them has diminished over time, thus reducing or eliminating the urgency to develop a plan. A species may have faced a high threat in one year but only a moderate or low one in subsequent years. After the FWS purchased part of the habitat of two endangered fishes in the Ash Meadows area of California and Nevada, as an example, the Service downgraded the threat from high in 1983, to moderate in 1986. Similar changes account for a fair number of the species without recovery plans. In contrast, the situation with some species has deteriorated so much that the development of recovery plans would be a misplaced effort.

Table 10.3 Percentage of Species with Recovery Plans by Level of Threat and Recovery Potential

Threat: Recovery potential:	High High	High Low	Moderate High	Moderate Low	Low High	Low Low
1978 (%)	96.1	86.0	84.8	75.7	95.5	100.0
(N)	26	43	46	37	22	6
1983 (%)	83.8	82.2	93.8	80.0	86.7	25.0
(N)	68	73	97	15	15	4
1986[a] (%)	61.6	73.6	76.1	66.7	66.7	50.0
(N)	112	110	117	24	21	2

Source: Compiled from data provided by the FWS.
[a]Includes species listed as endangered or threatened on 31 December 1988 and all plans approved through 31 December 1989, except those species subsequently delisted or for which a recovery priority ranking was not available.

The ivory-billed woodpecker provides an example. In the judgment of the FWS, the bird was a high-threat/low-recovery potential species in both 1983 and 1986. More important, a review of the species' status found little evidence that the woodpecker still exists in the United States.[18]

Considering these two possibilities still does not account for all the species with the greatest need for recovery plans. As an illustration, six species classified as having a high threat and a high recovery potential in both 1983 and 1986 were without recovery plans at the beginning of 1990.[19] All but one of the six species had been listed in 1979 or earlier. For another five species identified as having a high recovery potential in 1983, the lack of a recovery plan may have been especially disadvantageous. By 1986 the FWS acknowledged that the recovery potential for all five had deteriorated to the "low" ranking.[20] To underscore the plight of these species, three of them had also faced a high threat in the earlier year, thus making them candidates for immediate attention. Four of the five species are Hawaiian plants, each of which the Smithsonian Institution had identified in 1975 as in need of immediate recovery efforts. Even more ironic, one of these plants, *haplostachys haplostachya*, is a monotypic genus, thus giving it the highest of eighteen priority rankings in 1983.

Haplostachys haplostachya is not the only highly ranked monotypic genus without a recovery plan. There are few such monotypic genera, so comparison is difficult. Having noted this, several instances exist in which species and subspecies have plans developed before monotypic genera with the same level of threat and prospects for recovery. The monotypic genera affected are of particular interest because they reveal anomalies in the distribution of resources and effort.

At least one other anamoly is worth noting, namely the mixed pattern associated with the development of recovery plans for native species found in U.S. possessions. Recovery plans have been completed for all species listed before 1987 in Puerto Rico and the U.S. Virgin Islands. In contrast, at least through mid-1990, no recovery plans had been completed for species found on Guam or other U.S. possessions in the Western Pacific. If the many recent extinctions in this area are any indication, the remaining species are highly vulnerable to the same fate, including at least one monotypic genus.

Although some exceptions exist to the recovery priority scheme, they do tend to be few in number. In other words, the development of recovery plans parallels this priority system. This is also true for conflict-prone species. Of the twenty-one such species so identified in 1983, all but one had a recovery plan by 1987. Indeed, by the latter year the one exception faced only a low threat, and it no longer generated any conflict with a federal developmental activity. Eighty-nine species had been given the conflict label by 1986; slightly over three-quarters of these had recovery plans approved by the end of 1989.

Do Recovery Plans Make a Difference?

A recurring presumption is that recovery plans contribute to the preservation and recovery of listed species. The FWS prepares plans for species most likely to benefit from recovery efforts and the agency devotes considerable resources to the preparation of these plans. Plans identify what needs to be done, either to prevent extinction or, more desirably, to improve a species' prospects for complete recovery. To accomplish these objectives, a recovery plan includes strategies designed to reduce or eliminate threats to listed species. Once these strategies are implemented, the prospects for a species' recovery ought to improve, at least if the plan is well conceived. This was the message that Secretary of the Interior James Watt delivered to a congressional committee in 1983. Reviewing the Interior Department's accomplishments during his tenure, Watt pointed to "the vigorous implementation of the recovery planning process." Indeed, he declared attention to recovery plans to have been "a phenomenal success story. Our recovery program is helping us to take care of those species that are listed as threatened and endangered. And if we don't take care of these species, it doesn't do an awful lot of good just to list them."[21] Watt was surely correct, but his criterion for success was the number of recovery plans approved. The number of such plans represents a measure of administrative effort, not a consideration of the consequences of having the plans. A more appropriate measure involves improvements in the well-being of listed species. Do species with recovery plans find that their chances for recovery are enhanced and the threats facing them diminished?

It is possible to answer this question by examining changes in assessments of threat and recovery potential. These assessments are available for 1978, 1983, and 1986, as noted earlier, so the conventional wisdom about the effectiveness of recovery plans is readily tested. In brief, species with recovery plans should find their situation improved more so than species without recovery plans.

As the figures in table 10.4 clearly reveal, it is premature to conclude that recovery plans make much of a difference. The data indicate how the level of threat and recovery potential changed for species that had recovery plans approved between 1975 and 1982. For the majority of species in all three time periods, no change occurred. In other words, having a recovery plan did not alter the threat or the prospects for recovery for most species. In contrast, recovery plans did not prevent several species from encountering moderately worse threats or diminished opportunities for recovery. Among different taxa, mammals and birds were most likely to experience increasingly worse threats; fishes, reptiles, and plants were most likely to find their prospects for recovery diminished.

Perhaps more important, instead of the expected positive relation between recovery plans and improved status, just the opposite is evident. For species without recovery plans, the threat they faced decreased more than was the case for species with plans for each of the three time periods in table 10.4. In two of the time periods the relationship is the same for changes in recovery potential. Finally, when one considers species with plans approved over a longer time period, such as from 1975 through 1985, a similar pattern results. Here again,

Table 10.4   Recovery Plans and Changes in the Status of Listed Species (in percentages)

| Change from: | 1978–1983 | | 1983–1986 | | 1978–1986 | |
Recovery plan?	Yes	No	Yes	No	Yes	No
A. Threat[a]						
Significantly diminished	1.6	—	1.4	—	—	—
Moderately diminished	9.7	31.2	8.6	12.5	14.8	31.2
No change	64.5	43.8	77.1	82.5	55.7	43.8
Moderately worse	24.2	25.0	12.9	2.5	26.2	25.0
Significantly worse	—	—	—	2.5	3.3	—
Totals	100.0	100.0	100.0	100.0	100.0	100.0
(N)	62	16	70	40	61	16
Tau_c	−.09		−.08		−.11	
B. Recovery Potential[b]						
Improved	17.7	18.7	5.7	7.5	18.0	31.3
No change	71.0	68.8	81.4	80.0	64.0	56.2
Worse	11.3	12.5	12.9	12.5	18.0	12.6
Totals	100.0	100.0	100.0	100.0	100.0	100.0
(N)	62	16	70	40	61	16
Tau_c	.00		−.02		−.10	

Note: Includes recovery plans approved through 1982. Tau_c is a measure of association or correlation between two ordinal-level or rank-ordered variables. Its maximum values range from +1.00 to −1.00. High values indicate either a strong positive or strong negative relationship between variables. Values that approach zero indicate the absence of an association.

[a]There are three levels of threat: high, moderate, and low. Species are categorized according to how their threat changed during each time period. For example, a species facing a significantly diminished threat would be one that changed from a high to a low level of threat. A species facing a moderately diminished threat is one that changed from high to moderate or moderate to low. Changes in the opposite direction indicate a deteriorating situation.

[b]There are two levels of recovery potential: high and low. Species with improved prospects for recovery changed from low to high potential or vice versa for species with diminished prospects for recovery.

recovery plans do not seem to provide any noticeable benefits. In fact, no species with a recovery plan had been fully recovered, at least through early 1990.

These findings are contrary to expectations and require some consideration of possible explanations. Most obviously, some plans are developed and implemented too late to be of any long-term value. A problem with many other plans is that they establish recovery as a goal, but then do not indicate how success is to be judged.[22] Other plans similarly do not provide estimates of costs or time frames for recovery. The best and most plausible explanation is the limited relation between a plan's approval and its implementation. As one observer of the recovery process has observed, "recovery plans, like prescriptions for medicine, are pieces of paper; unless one buys the medicine prescribed, recovery will not result."[23] Withholding proper medicines from human patients would be deemed indefensible, but this is a common occurrence for listed species. Many species have recovery plans, but many tasks in these plans have yet to be implemented. In one of its 1980 planning documents, as an illustration, the FWS established as a goal the full implementation of only 15 percent of its recovery actions and the partial implementation of another 75 percent by 1985.[24]

This goal was not achieved primarily because of the limited availability of money. The FWS suffers from an uncertain dependence on congressional appropriations. The number of recovery plans increased significantly in the 1980s, but congressional support for the implementation of these plans did not. From the beginning of fiscal year 1982 through the end of fiscal year 1987, a six-year period, the number of species with recovery plans increased sixfold from 42 to 255. Over the same period, congressional appropriations for FWS recovery efforts increased by only 22 percent, to just under $6.4 million. Accounting for inflation, which was relatively high in the early 1980s, the small increase becomes even less meaningful.

Once the FWS receives its appropriations for recovery efforts, choices must be made about how to spend the money. The FWS does not revise its planned expenditures for recovery activities every time a new plan is finished, and it is unreasonable for the FWS to do so. Species with "older" plans have an advantage. If recovery activities have been initiated for the species, the activities can require expenditures for several years. It is unlikely that recovery actions for one species will be terminated in order to provide funds for another species and its recovery plan except in unusual circumstances. Few administrative agencies, including the FWS, can afford to ignore sunk costs and past expenditures. Once a recovery task begins it makes little sense to halt it, particularly when the action prevents extinction. Prematurely stopping a long-term recovery effort risks losing everything that has been gained. Furthermore, resources and expertise devoted to bats are not easily transferred to recovery efforts for endangered clams or butterflies.

The FWS must also choose between spending large amounts for a few species or distributing the same resources among many species that are less costly to recover. The Service's stated intent is to distribute its recovery resources equitably among all listed species.[25] At least one exception exists. The FWS attempts to take the emergency measures necessary to stabilize the status of all high-threat species or to prevent their extinction. If stability rather than recovery is the goal, then the data in table 10.4 point to many successes.

For the opponents of extinction this is a reasonable approach; it is also an expensive but not necessarily an economically efficient one. Measures designed to save a species from extinction or irreversible decline are often costly and cannot be delayed, at least if the goal is to avoid extinction. According to FWS estimates for 1985, the agency would need $20 million above existing appropriations over five years not to recover, but just to reduce the threat of extinction or significant decline for about sixty species whose status is critical.[26] This would include over $2.3 million for the southern sea otter, $2 million for three Colorado River fishes, $1.8 million for the black-footed ferret, and slightly less than $1 million for the whooping crane. Spending all this money would not ensure the species' recovery. Additional funds beyond the five years would still be needed for many of the sixty species.

Even nonemergency measures can be expensive. In making projections about the estimated costs of recovering the blunt-nosed leopard lizard, a species listed as endangered since 1967, one FWS specialist concluded that more than $41 million would be needed over a ten-year period.[27] This is nearly twenty times as much as the FWS had spent on the species in the previous five years.

These examples offer some evidence that economic efficiency is difficult to achieve. Such efficiency would require the FWS to maximize the number of species recovered. Money would be spent on species with the least costly requirements before recovering more expensive species. The FWS has not adopted this approach and, consequently, has delayed the recovery of several species with a high potential for recovery. In one of its planning exercises, for example, the FWS estimated that the Borax Lake chub, an endangered fish in Oregon, could be recovered for as little as $8,000.[28] Other species would require higher expenditures, but the exercise concluded that sixteen so-called threshold species are recoverable within five years at a cost of less than $1 million. This analysis of expenditures suggests that the FWS believes it to be more desirable to do what it can to prevent the extinction of a species than it is to recover many other species in somewhat less danger. This is a controversial approach that subjects the Service to criticism from a Congress that wants multiple demonstrations of success. The Senate Committee on Environment and Public Works has made clear its preference for attention to species that are closest to recovery rather than to species closest to extinction.[29]

Even if its errs in its emphasis, the FWS does not do so consistently. Although the preference of the FWS is to divide its recovery resources equitably and to avoid extinction, there is considerable evidence that the FWS does not follow its own guidelines. Nearly 45 percent of the money available for the development and implementation of recovery plans between October 1981 and September 1986 was devoted to just twelve animals, or about 5 percent of all listed U.S. species.[30] In contrast to the favored birds, mammals, and sea turtles, little money was devoted to the recovery of plants, insects, mollusks, or crustaceans.

The bald eagle was one of the favored species. Compared with most other listed species, the bald eagle faces only a modest threat to its continued existence. Writing in the mid-1970s, one FWS biologist said that the eagle is a species "in relatively fair condition," whose future is almost certain.[31] The first FWS recovery priority system, published in 1978, gave the bald eagle the lowest of twelve rankings, which was reserved for subspecies with a low threat. More recently, the bald eagle has been ranked as a species with a low threat and a high recovery potential. This means that the bald eagle ranks below more than 90 percent of all listed species in terms of recovery needs. Despite its low rank, the bald eagle has been the beneficiary of considerable publicity and enhanced law enforcement activities. Millions of dollars have been spent on research and recovery, federal grants to the states, and on land acquisition with money from the LWCF. All these actions (and the banning of DDT in 1972) have allowed the eagle's population to soar everywhere it is found in the United States. Asked to explain why the low threat eagle received so much financial attention, FWS officials acknowledged "that the need to achieve a positive public perception of the program sometimes drives the agency to devote extra attention to species with high 'public appeal,' such as the bald eagle."[32] This justification obviously contradicts the previously stated position that nonbiological considerations are irrelevant in decisions about recovery.

Variables other than internal resources provide explanations for the ineffectiveness of many recovery plans. Good intentions are not enough to overcome disrupted habitats or a poor understanding of the threats that place species in jeopardy. Even when understanding is present, many actions have a low probability of success. Furthermore, most listed species do not reside on lands that the FWS manages, so the agency is dependent on the cooperation of scores of state and federal agencies as well as on thousands of private landowners.

This dependence can easily work to the disadvantage of endangered species. Efforts to protect the Socorro isopod, a small crustacean, are hampered because the species' only natural habitat is on private land, and the owner is unsympathetic to the isopod's predicament. The prospects for the endangered red-cockaded woodpecker did not improve when a disgruntled land developer in Florida shot some of the birds and destroyed nesting sites because he believed concern for the woodpeckers was unduly delaying his application for a building permit.

The situation for listed plants is even more remarkable. It is unlawful to harm an endangered animal intentionally, but it was entirely legal to pick or destroy an endangered plant on private property prior to passage of the Endangered Species Act Amendments of 1988. Damaging or destroying a listed plant was illegal only when it occurred on federal land *and* when the plant was removed from this land. One consequence was that some people could legally encourage the private collection of rare species, and it did not help the recovery of an endangered cactus when a magazine published directions to the species' only known habitat in New Mexico.[33] The Navasota ladies' tresses is considered to be the rarest orchid in North America, but this did not deter one landowner from destroying some of the plants so that he could make his place of business more visible.

Public agencies may offer symbolic support for the protection of endangered species, but this does not guarantee their ability or willingness to implement recovery actions or to commit scarce resources to the protection of these species. As a GAO report noted, recovery plans are intended to achieve biological goals without consideration of political or socioeconomic factors. The result, said the report, was

FWS approval of recovery plans "that were not readily attainable because they conflicted with the views, interests, and responsibilities of participating individuals and agencies, and/or coordinating agencies did not have the funds to implement them."[34] As Hawaii's experience with the palila demonstrates, sympathetic responses to recommended recovery actions are not always assured.

Other examples demonstrate this point. When the FWS asked the army if Mexican wolves could be reintroduced into its White Sands Missile Range in New Mexico, the army said no. State wildlife agencies in Texas, Arizona, and New Mexico also failed to recommend sites for reintroduction. In Texas, the state legislature passed a law prohibiting the wolves' reintroduction within the state.

The FWS has long emphasized its dependence on the states. This relationship became institutionalized with the passage of the Endangered Species Act of 1973. The law's Section 6 provides for the establishment of cooperative agreements between the states and the FWS and NMFS. Once an agreement is signed, management of resident species of listed animals as well as some responsibility for their recovery is largely given to the state. The 1978 amendments to the law authorized similar cooperative agreements for the protection of plants.

The FWS hoped that states would quickly sign agreements. Lynn Greenwalt, FWS director in 1975, expressed these hopes when he stressed that "without the support and contribution of the States, the effectiveness of this act has very little future, in my judgment."[35] To encourage the states to enter into agreements, Section 6 authorizes matching grants to state agencies. Each state must meet certain requirements before it is eligible for federal grants. Indeed, this federal support is one of the main reasons that states are willing to establish agreements. The problem, however, is that both congressional and presidential support for the grant program has been erratic.

Congress did not provide any money for state grants until two years after passage of the 1973 law. For fiscal year 1977, the FWS requested $4 million for state grants, but Interior recommended only half that amount to the OMB. In turn, the OMB did not ask for any money from Congress. For fiscal years 1979 through 1981, Congress provided slightly less than $12 million to the FWS for state grants. For the next three fiscal years, however, President Reagan did not request anything. Congress ignored the president in two of the three years and appropriated $4 million. The president requested money for the program in 1985 and 1986, but not a penny for fiscal years 1987 or 1988.

Some money was available in 1987, but many states complained that it was entirely inadequate to meet the growing need for assistance. The average grant had been almost $200,000 per agreement in 1977.[36] Ten years later the average was $53,000 per agreement, and several states with agreements did not receive any money. So frustrating has the process been for some states that they have simply declined to ask for any money with the consequence that they curtail efforts to protect listed species.

When money is available, negotiations between the states and the FWS determine the projects to be funded. This approach does not ensure that the species most in need of attention receive it. Through the first six years of funding, almost 70 percent of the money devoted to federally listed species went to birds and mammals, even though they accounted for less than 40 percent of all species listed.[37] More money was spent on the bald eagle than on all other species except one, the peregrine falcon.

The federal government's inconsistent support for the grant program provides an explanation for some states' reluctance or lack of desire to negotiate an agreement. Louisiana and Alabama, both of which have many endangered species, do not have agreements for either plants or animals. Other states do have agreements with the FWS, but it took many years to negotiate them. States could enter into agreements for plants beginning in early 1979, but only nineteen states and Puerto Rico had such agreements by the end of 1985. Four years later there were still more than a dozen states that did not have cooperative agreements for plants. Despite the dramatic need for recovery activities in Hawaii, the FWS was unable to provide any Section 6 grants to the state until 1984, when an agreement was achieved.

Not only must the FWS contend with some states' apathy and apparent hostility, but it must also deal with similar attitudes within Interior, as the experiences with the northern Rocky Mountain wolves and the San Diego mesa mint reveal.

The Wolves

Northern Rocky Mountain wolves represent one of three populations of gray wolves for which the FWS has developed a recovery plan. An initial plan approved in 1980 soon became outdated. Work to update it began a few years later. When the revised plan was finished in late 1986, one recommended strategy was the experimental reintroduction of a small number of wolves to Idaho, Montana, and Yellowstone National Park. William Penn Mott, Jr., director of the NPS, enthusiastically supported the reintroduction into Yellowstone. The wolf, he said, "is a symbol of the West. For people to be able to hear a wolf howl is going to be a very exciting opportunity."[38] In spite of this enthusiasm, the proposal generated heated opposition from hunters, ranchers, Wyoming's small but powerful congressional delegation, and from many state legislators, one of whom complained that having wolves in national parks is like having "cockroaches in your attic." The pressure on Interior was intense and effective. After having approved the recovery plan, FWS director Frank Dunkle termed the proposed reintroduction "foolhardy." He also pledged to erect every administrative roadblock imaginable if legislation requiring reintroduction of wolves into Yellowstone National Park was approved.[39] Faced with congressional opposition and Dunkle's change of heart, Mott announced a delay in reintroduction until Wyoming's three congressmen approved. This approval was not easily obtained. Largely because of his opposition to the reintroduction program, Senator Alan K. Simpson (R.–Wyoming) delayed for three years, from 1985 to 1988, the Senate's consideration of a bill to reauthorize the Endangered Species Act.

The San Diego Mesa Mint

Listed as endangered since 1978, the mesa mint grows only along the edge and bottom of dry or drying shallow spring rain, or vernal, pools north of San Diego, California. Some of the mint's specialized habitat coincided with land to be developed for an industrial park. Vernal pools were involved, so potential developers had to obtain a permit from the Army Corps of Engineers, in accordance with its responsibilities for certain wetlands under the Clean Water Act.

One corporation complained to the assistant secretary of the interior for fish and wildlife and parks that concern for the mints was delaying construction and hindering efforts to comply with the permit requirement of the Corps. At a meeting to discuss how the complaint could be resolved, the assistant secretary asked why the "weed" could not be transplanted and maintained through irrigation.[40] One of Interior's lawyers was far more direct in expressing his feelings. The lawyer, who one might expect to be sympathetic to the plant's legally mandated protection, suggested that the corporation should have fired its legal counsel for failing to advise the destruction of the mints on the corporation's property before applying to the Corps for a permit! With this administrative hostility, one can easily imagine that the recovery plan eventually approved for the mint was written to ensure its acceptance within Interior.

The Palos Verdes Blue Butterfly

Even with sympathetic officials, ineffective communication can foil recovery actions, as the Palos Verdes blue butterfly illustrates. When the butterfly was listed as endangered in 1980, only three years after it was described scientifically, the FWS also designated three areas as critical habitat. All three areas are located in Ranchos Palos Verdes, an affluent suburb of Los Angeles. Soon after the butterfly's listing, the FWS contracted with a lepidopterist, a specialist on butterflies, to develop a recovery plan. The draft recovery plan identified the major threats to the species as agriculture, residential growth, weed abatement, and the periodic tilling of fields to create temporary firebreaks.[41] These activities destroy the locoweed that provides the species' only source of food.

Local public officials expressed their willingness to assist in the preservation of the butterflies, but this did not produce the desired results. Before final approval of the recovery plan, all the host plants at the second largest colony were destroyed through tilling.[42] What had once been the habitat and breeding ground of a unique species had now become a baseball field. After this episode it was not surprising for the recovery plan to emphasize that: "City and fire district officials need to be informed about recreation and fire preventive activities which are compatible with preservation of the butterfly and its habitat."[43] Within a few months of the recovery plan's completion, the FWS wrote to city officials in Ranchos Palos Verdes, reminding them that two butterfly habitats in the city should be left undisturbed and emphasizing that the sites would provide "the best chance for finding butterflies."[44] Unfortunately for both the FWS and the butterflies, the letter arrived soon after both sites had been cleared of vegetation, including the locoweed. Subsequent searches of previously known habitats found no eggs, caterpillars, or butterflies. The only suitable habitat had been irreparably destroyed, and the Palos Verdes blue butterfly is now believed to be extinct.

In their defense, officials in Ranchos Palos Verdes explained that they have never been told how to protect the butterflies. "I've never been sure what our responsibility is," said the city's environmental services director. As he added, the FWS keeps "telling us to protect the habitat, but we've never been told how."[45]

The California Condor

Uncertainty about how to preserve endangered species is not limited to elected officials. Every recovery plan recommends actions that are *believed* to be necessary to bring about recovery. Whatever the intentions, there is no guarantee either that the actions will lead to success or that they are even appropriate. In its plan for the California condor, one of the most studied of all endangered species, the plan's authors pointed out that:

> the greatest problem in condor management is the lack of precise data concerning the needs of the species. Despite considerable research, it is sometimes impossible to concretely justify recommendations or give positive answers to questions regarding the impact of certain actions and developments. For instance, it is not always possible to say whether or not a certain oil development, logging operation, public use facility, or new roadway will definitely impair the condor's chances for survival.[46]

Having said this, the recovery team attempted to identify all of the condor's needs and then proposed a comprehensive strategy designed to meet these needs. As the team stressed, implementation of all items in the plan would offer the best chance for the condor's survival.

Over the next dozen years, Interior, the USFS, the California Fish and Game Commission, and the National Audubon Society would spend millions of dollars and devote thousands of hours to implementing the plan. Several areas were designated as critical habitat, a Condor Research Center was created, and Congress provided several special appropriations to expedite the recovery program. The FWS, USFS, The Nature Conservancy, and the California Wildlife Conservation Board all purchased thousands of acres of habitat. Perhaps more than nearly all other endangered species, efforts to recover the condors benefitted from constant media attention, with television specials, books detailing the condor's plight, and hundreds of news-paper and magazine articles.[47] The FWS could not have asked for more in terms of effort, public awareness, or good intentions. These were not enough. The number of condors in the wild declined, from an estimated fifty to sixty in 1968, to twenty-five to thirty in 1978, to about twenty in 1983. By 1986, only two males remained in the wild, but both were captured in early 1987 as part of captive breeding programs in zoos in San Diego and Los Angeles.[48]

The species will survive for some years in the zoos, but the notion that captive condors can be reintroduced into their natural habitat is questionable. Much of their habitat is being developed, both zoos have doubts about the reintroduction program, and the captives may be too tame to scavenge and survive in the wild. This is certainly so for the condors in Los Angeles. Five of them were taken out of iso-lation, placed in regular cages, and fed by human handlers. After all its efforts, the FWS had failed to preserve the species in its natural habitat and still had not been able to identify the primary causes of the condor's decline.

Tallying Results

Successful efforts to recover endangered and threatened species would produce a long list of species whose existence is no longer in doubt. As noted earlier, however, no recovery plan had brought about the formal recovery of a single species by early 1990. Four species had been recovered before then, but none had a recovery plan. Three bird species on the Palau Islands, near Guam in the western Pacific, were delisted in 1985. The species' "recovery" was due to the discovery of additional birds rather than to any recovery efforts. The fourth species, the American alligator, has recovered, but many people dis-agreed that it was ever in jeopardy. In addition to these species, eight other species and two populations had their status changed from endangered to threatened.[49] Only three of the eight species had re-covery plans before the change in status occurred. Moreover, the FWS changed the status of several of the species in order to allow regu-lated hunting or fishing, actions that endangered status prohibits but which threatened status permits through special rules.

In contrast, six native species—four fishes, one bird, and one mus-sel—have formally been declared extinct.[50] Many more species are considered to be extinct, but have not yet been formally removed from the list of endangered species.[51] The FWS publicly announced the death of the last dusky seaside sparrow. The Palos Verdes blue butterfly's habitat has been destroyed, the Mexican wolf and the California condor are not surviving in their natural habitats. The only known population of black-footed ferrets is in captivity, the ivory-billed woodpecker has not been seen in the United States since the 1950s, and the continued existence of many other listed species is in question. As a result of several status surveys, the NMFS concluded in late 1984 that the Caribbean monk seal is extinct and should be removed from the list of endangered species.[52] No action has been taken to do so.

Others species survive in such low numbers that the premature death of two or three individuals would ensure the species' extinction. For example, when a plant, Kearney's blue star, was listed in 1989, botanists knew of only eight individuals. Similarly, a 1981 survey of the Kauai O'o located only two birds; the Kauai 'akialoa and the Kauai nukupu'u, two Hawaiian forest birds, have been sighted only a few times this century.

Some species have a larger number of survivors, but not enough to avoid genetic inbreeding. So few woodland caribou remain in Idaho and Washington that "the premature loss of a single individual could be disastrous to the herd."[53] For still other species, their habitat is confined to a single, highly vulnerable location. Vandalism, a natural disaster, an unexpected disturbance of a habitat, illegal hunting or collecting, or the spill of a toxic chemical could cause the demise of these species. Altogether the FWS estimates that scores of listed species are unlikely ever to be fully recovered, regardless of effort. Finally, many native species have become extinct before they could be listed.

Conclusions

In attempting to return listed species to better times both the FWS and the NMFS find themselves significantly disadvantaged. Neither agency can do much to prevent a species from becoming endangered. Most of the agencies' patients are already desperately in need of help when they are listed, so the number and range of effective remedies is automatically limited. Indeed, when habitats are destroyed or altered, effective solutions may not be available. Resources are inadequate, and the necessary cooperation is not always assured. The FWS and the NMFS can ask other federal agencies for assistance and hope that private citizens will provide it, but often what needs to be done is far from certain, just as is the continued existence of hundreds of endangered species.

D epending on one's hopes and expectations, efforts to protect endangered species in the United States are either a modest success or a massive failure. Since their modest beginnings with passage of the Endangered Species Preservation Act of 1966, these efforts have since grown considerably in size, authority, and responsibility. Hundreds of native species are now formally listed as either endangered or threatened, scores have had critical habitats designated, and many have recovery plans. Furthermore, no federal agency can legally ignore its obligation to conserve listed species and their habitats. Programs in most states supplement these federal arrangements, and public opinion strongly supports the protection of endangered species, at least in the abstract. All of these features combine to produce what is arguably one of the world's best financed, most comprehensive, and organizationally mature programs to protect biological diversity.

Despite this acclamation, the program can point to few successes, at least when measured against its statutory goal. Only a handful of listed species have been recovered after more than two decades of effort and expenditure. Of these successes, few are due to program-related activities. During the program's life, in contrast, a larger number of species has become extinct, and the number of species in need of protection increases not only regularly but well beyond the government's capacity or willingness to respond. In fact, some species become extinct even before they have a chance to be listed. Being listed does not guarantee survival or recovery and, frequently, not even adequate protection from the causes of endangerment. When all is considered, much remains to be done, and the task is far from complete. The program can claim victory in a few skirmishes, but it is losing the larger and more important battle.

There are many reasons for this situation, as the previous chapters have shown. Among the explanations are both faulty implementation and a faulty impact model. Failures in the first category are easily identified, whether the topic is listing, protection of habitats, consultation, or the implementation of recovery plans. The FWS has the major responsibility for protection of endangered species, so it is a fitting target for condemnation. The FWS deserves a share of the blame, but this blame must be put in context. Congress has set a high and possibly unachievable goal, but it usually fails to match this goal with either sufficient resources or political support. However well intentioned, Congress also frustrates success because of frequent legislative changes. The primary responsibility of the FWS in regard to the Endangered Species Act is to recover species on the brink of extinction. The FWS is often obligated, however, to commit its limited resources and expertise to subsidiary and secondary goals such as the frequent drafting and redrafting of rules and regulations and the preparation and conduct of public meetings and hearings. These activities may be administratively and procedurally desirable, but they do little to protect biological diversity.

The Service's location in the administrative hierarchy further hampers its efforts. Buried deep in the bowels of Interior, the FWS finds that its priorities are not necessarily those of its superiors. The same kind of relationship likewise typifies its endangered species program, which is only one of many areas for which the Service has responsibility.

The Service has many constituencies, but primary among these are hunters and fishermen who remain content only as long as the Service provides victims in the form of fish and fowl. This interest in wildlife seems to make the FWS a suitable location for endangered species activities, but this location does not ensure a sympathetic response to the problem. Protection of endangered species often creates conflict with other activities in the FWS, and the program has never been the Service's favorite child, at least if the frequent reorganizations of the program serve as any indication. When Interior first became involved with endangered species in the early 1960s, an initial goal was the creation of a single office or section to coordinate responses to the problems of extinction and endangerment. With the creation of the OES, this goal was largely achieved. By the 1980s, however, the quest for improved administrative efficiency led to major changes in structure and finally to the demise of the OES. Its functions remain, but they are now divided and subsumed in a Division of Endangered Species and Habitat Conservation, which was created in late 1987. Furthermore, what was once a centralized effort directed from OES headquarters is now a demoralized and widely decentralized program with only a skeletal staff in Washington, D.C.[1]

Moreover, as recent years have shown, endangered species are more likely to embroil the FWS in controversy than to envelop it in favorable publicity. This is neither a recipe for administrative or biological success nor a desirable way to preface appeals for larger budgets.

The limited political clout of the endangered species program further reveals itself in FWS interactions with the partners it gained in a congressionally forced marriage. The FWS shares responsibility for the administration of the Endangered Species Act with the Departments of Commerce and Agriculture and for the protection of endangered species with many state and federal agencies. This arrangement is often one of meaningless form rather than of meaningful substance. Commerce has management responsibility for certain marine species, but this responsibility appears to be an unimportant and unwanted intrusion. No endangered or threatened marine species have been recovered because of Commerce's efforts, which have been minimal. The NMFS has listed species only when it has been forced or pressured to do so, and nearly the same can be said about its designation of critical habitats and its development and implementation of recovery plans.

The apathy of the NMFS is well recognized and regularly criticized, but little seems to change.[2] In 1979 Congress required the NMFS to establish a ranking system for species in need of listing and a priority system for developing recovery plans.[3] Nine years later neither had been prepared, but this did not seem to bother either the Senate or the House. Their review of the Endangered Species Act in 1987 omitted any mention of the NMFS's failure to comply with the law.

The Endangered Species Act also assigns a small role to Agriculture's Animal and Plant Health Inspection Service (APHIS). Its role is to regulate imports and exports of listed plant species. International trade of such species is quite large, but the response from APHIS has been negligible. Between late 1978, when it first gained responsibility for monitoring this trade, and late 1987, the APHIS had referred only three suspected violations of the law to the Department of Justice.[4] Over the same period, the FWS had made over one thousand referrals regarding illegal trade in protected wildlife species.

The three agencies primarily responsible for implementation have a mixed commitment to success, and the same assessment is appropriate for other federal agencies as well. These agencies are obligated "to take such steps as may be appropriate" to conserve listed species and their habitats. Whatever the congressional intent may have been, interpretation of this obligation is left to the self-defined intentions of individual agencies as they pursue policies often unrelated to the protection of biological diversity. Furthermore, the FWS and NMFS issue regulations that govern such areas as interagency consultation under Section 7, but initiation of consultation is left to the project agency, not the FWS or NMFS. Far more important, once a consultation leads to a jeopardy opinion, it is the project agency that accepts or rejects the recommended alternatives to its proposed action. Agencies are likely to accept these alternatives in some form, but this does not ensure either their eventual or effective application, as the EPA experience with pesticide labels well demonstrates.

The FWS also finds itself in a "marital" arrangement with the states and territories, the places where listed species are found. The various endangered species laws envisage a cooperative relation between the states and the FWS or the NMFS, but such cooperation is not always forthcoming. Federal grants are supposed to create incentives for the establishment and continued implementation of cooperative agreements between the two federal agencies and the states. Irregular and often parsimonious funding has, however, probably angered and alienated many states that have agreements. Even when disagreement about the availability of federal funds does not exist, relations between the FWS and the states still find occasion to be strained. Federal efforts to protect species often conflict with state and private efforts to develop *and* conserve (and, sometimes, just to develop). Relevant examples here include the attempted designation of a critical habitat for grizzly bears, the designation of such a habitat for the palila and monk seal, and the proposed reintroduction of wolves into several western states. In other instances some state agencies are likely to believe that the federal government is not sufficiently assertive in its protection of vulnerable species.

Just as faulty implementation can be attributed to the program, so also can a faulty impact model. At a minimum, an effective impact model must ensure that its component actions are directed at a problem's source and that the action agency knows how much attention to devote to each of these actions. Endangerment has many causes in the United States, but chief among these is the disruption or destruction of habitats, an action now not subject to effective control or prevention because of the current impact model. Until government policy shifts to one of prevention and away from the current policy of reaction, endangerment and extinction will be increasingly common. The species that vanish will include many the nation can ill afford to lose. In short, the current impact model is incomplete.

This model is further deficient, at least if the FWS's use of it provides any lessons. The model includes the development of regulations, listing, the designation of critical habitats, consultation with federal project agencies, and the implementation of recovery plans. The relative importance of each of these tasks has varied, and the FWS has shifted emphasis from one aspect to another over the last fifteen years, not always with biological justification or in accordance with the agency's own priority systems. In the early 1970s, as an illustration, the FWS emphasized a need to develop a comprehensive set of regulations to justify aggressive implementation. By the early 1980s, however, the advent of the Reagan presidency and several changes in the Endangered Species Act caused considerable delay in the issuance of regulations. Likewise, at one time there was an emphasis on listing, with the assumption that more was better. Ten of thousands of native species would soon be listed, or so the FWS indicated in the mid-1970s. Increasingly burdensome procedural requirements, added in 1978, and opposition to listing large numbers of additional species made listing a less attractive exercise in the 1980s.

The designation of critical habitats, first allowed by the Endangered Species Act of 1973, was initially viewed as an effective way to protect endangered species and their habitats. Indeed, designation was deemed to be so vital that the Departments of Interior and Commerce once received a presidential directive to expedite the designation of critical habitats. The FWS consequently announced its intention to determine critical habitats for nearly all endangered and threatened species, and for many of these as quickly as possible.

Within a few years of this announcement serious disagreement existed within the FWS about the utility of designating critical habitats, and the agency's own system of priorities for designating habitats was ignored. By the mid-1980s decisions to designate habitats frequently depended on the personal preferences of the agency scientist with primary responsibility for shepherding a listing proposal through the bureaucratic maze. Those opposed to designation appeared to gain the upper hand in the late 1980s; by then the designation of critical habitats was the rare exception rather than the usual rule. As an illustration, there were more critical habitats designated in 1976 and 1977 than in the last four years of the 1980s, despite the significantly larger number of listed species in the latter period.

For the NMFS, designation has never been a popular activity. The agency acknowledged in 1978 that many of the species for which it has responsibility could face extinction "in the near future" unless critical habitats were designated, but this judgment generated almost nothing in the way of concrete actions.[5] Of the five NMFS species facing the highest threat and the greatest danger of extinction in 1978, only one, the Hawaiian monk seal, has had a critical habitat designated, and not necessarily because the agency wanted to do so.

The FWS initially claimed that it would be assertive in the use of the consultation process, that jeopardy opinions would be common, and that many projects would be halted or altered substantially in order to protect listed species. These claims bore fruit in a number of instances, most notably with the Tellico Dam and the snail darter. As a result of this clash and another involving whooping cranes, Congress added an elaborate exemption process in 1978. This process, which relies on an Endangered Species Committee composed of high-ranking state and federal officials, is to be used whenever a federal agency believes that a jeopardy opinion would preclude completion of one of its projects. The clear expectation was that as the number of listed species increased, so also would the number of "irresolvable" conflicts, especially if the FWS remained as combative as it had been in the snail darter controversy. In fact, however, the number of consultations is far less than the FWS once predicted, and rarely do these lead to jeopardy opinions. Among all jeopardy opinions, few have ever halted a federal project, and none have required the services of the Endangered Species Committee, at least since the cases with the snail darters and the whooping cranes. The absence of seemingly irresolvable conflicts could be due to a decline in the number of federal projects adversely affecting listed species (an unlikely prospect), a greater sensitivity on the part of federal agencies to listed species (a possible but not completely satisfactory explanation), or to the increased desire of the FWS to avoid conflict with its sister agencies (a distinct likelihood, especially in the 1980s). The so-called Windy Gap decisions offer evidence of the latter scenario. In sum, although the consultation process is useful, it is far less important than the FWS once envisaged it would be. By an overwhelming margin, most consultations are routine and informal, and only infrequently do they require an in-depth biological investigation of a listed species' situation.

After emphasizing the importance of the consultation process in the late 1970s, the FWS focus again shifted in the 1980s to the development of recovery plans at the expense of listing additional species. The number of completed plans increased rapidly during the early 1980s; by the late 1980s, however, the number of new and revised plans slowed considerably. The purpose in preparing plans is to save endangered species, because merely listing a species is usually insufficient to bring about its recovery. Remedial actions are always necessary, thus the need for recovery plans that specify the particular steps that should be taken to improve the prospects for a species' return to viable numbers. Despite this explanation, as more plans were approved in the early 1980s, the gap widened between the number of plans approved and the number of plans either fully or partially implemented. The FWS finds itself prescribing medications that it cannot or does not want to buy. Even among plans that have been implemented, Congress has found reason for criticism, either because they are incomplete or because many plans fail to provide indicators of success. In other words, many recovery plans themselves suffer from faulty impact models. A further problem arises because the FWS can only recommend recovery strategies. The Service cannot require or guarantee cooperation from the public or other government agencies in the implementation of these strategies.

Fully implemented recovery plans are desirable, but as with many other program elements, the FWS is faced with undesirable choices brought about because of limited budgets. Devoting more money to recovery means less attention to listing, consultation, or the development of additional recovery plans. When one of these latter tasks is given priority, recovery efforts will suffer, at least so long as congressional appropriations and current patterns of implementation remain relatively static.

The best evidence for this conclusion is the limited number of species that have been recovered. Not all species have recovery plans, but among species that do, their plans do not guarantee recovery and occasionally do not even contribute to an improvement in their situation, at least if the analysis in the previous chapter is correct. Part of the explanation for this is due to the Service's decision to pursue a goal that differs from the one Congress has mandated, namely the recovery of listed species. The FWS does not reject this goal but believes it is more desirable, given the agency's circumstances and the plight of those entrusted to its care, to concentrate its limited resources on the species on the edge of extinction rather than on those closest to recovery. This preference maximizes the number of species that can be saved over the long term but minimizes the number that are recovered over the short term. The FWS approach is ecologically defensible, but not politically astute. Congress repeatedly expresses its preference for short-term results in the form of recovered species. Such successes contribute to an aura of administrative accomplishment, a conclusion that resources are used wisely and effectively, and a belief that further or even enhanced support is merited. The FWS risks the loss of each of these advantages by focusing on the prevention of extinction rather than on recovery. No administrative agency finds it easy to capture credit or attention for preventing an event, such as extinction, that does not occur.

Perhaps more important, the lack of many recoveries suggests that the causes of endangerment are beyond solution or that the costs of recovery are higher than what Congress is willing to provide, either financially or in terms of more effective responses to the causes of endangerment. Here the research of Anthony Downs and John Kingdon may be instructive.[6] They observe that attention to a problem begins to lag when public officials realize that their policies have not produced the anticipated results or when these officials begin to appreciate the costs of success. As Kingdon reasons, "As people become impressed with the sacrifices, dislocations, and costs to be borne, they lose their enthusiasm for addressing the problem."[7] He adds that enthusiasm also subsides when a problem loses its novelty, even though the problem remains important and continues to deserve attention.

The issue of endangered species has not moved off the federal government's agenda, but neither has it moved to center stage where many scientists believe it should be alongside other environmental problems like toxic waste and air pollution. Indeed, recent experience suggests that the issue of extinction is not likely to rise in prominence in the near future. What is possible, however, is a gradual decline in attention to the issue unless some kind of episode occurs that is recognized as an ecological disaster or until the American public becomes better informed and more concerned than it is now about the causes of endangerment and the consequences of extinction.

The FWS has undertaken many efforts to publicize the plight of endangered species, but after twenty years there is scant evidence to show that these efforts have created the kind of broad-based public support or awareness that could contribute to the success of an effective program for endangered species. Concern for some species like pandas, whales, gorillas, and elephants does exist and is commendable, but this concern slights the far more numerous but less visible endangered plants, reptiles, insects, and other invertebrates that provide the foundation for ecological stability. Under the best of circumstances, advancing public knowledge and awareness of these species will be difficult, as hundreds if not thousands of concerned scientists have already discovered. Some people have suggested that the economic value of endangered species be highlighted. Others just as readily reject this idea because most listed species have no palpable or apparent economic value. Furthermore, assigning such value could promote the demise of a species if it finds itself in competition with an appealing project assigned a higher value.

Whatever the case, people have to find reasons to change their behavior and to alter the activities that contribute to the extinction of so many species. One can hypothesize that if the current regulatory approach to preservation continues to be the chosen path, then Congress will not require more from the public than it demands or is willing to tolerate. As Harold and Margaret Sprout have generalized about Americans, "it is difficult to discern much willingness to pay a stiff price in money or inconvenience for [environmental] protection."[8] For endangered species, the evidence for this comes in the form of things Congress could easily do, yet so far has chosen not to do. If Congress wanted to strengthen the existing endangered species laws while maintaining the current framework, for example, lawmakers could do several things. First, and most obvious, Congress could increase dramatically expenditures for all facets of the program, from listing, consultation, and recovery to state grants and land acquisition. The amount the federal government now spends each year to protect endangered species is a paltry fraction of the cost of just one of the military's latest jet fighters. Although the 1980s were supposedly a time of tight budgets and limited resources, Congress was able to provide money for nearly all the programs it considered deserving, however wasteful or dubious their merits.[9] Exceptions exist, but when Congress and the president believe a program to be vital, it gets funded at adequate levels and frequently well beyond these levels. Less favored activities do not find themselves to be the beneficiaries of congressional munificence, and such is the case for endangered species. The FWS is now faced with many undesirable choices among vulnerable species due to lack of money. In the absence of adequate resources, including staff, many deserving species do not get listed while others wait years for recovery plans only to suffer further because the FWS cannot afford to implement them. Additional funding will not ensure the survival of all listed species, but many more would have a far better chance than now exists.

Second, the increasing emphasis on process at the expense of substance could be reversed. Consider that the FWS has withdrawn proposed listings for about four times as many native species as it has actually listed over the last twenty years. The FWS and the Smithsonian Institution said that about eighteen hundred plants required the benefits of listing in 1976, when they were first proposed. Over the next two years the threats facing most of these plants in no way diminished, but procedural changes added to the Endangered Species Act in 1978 caused the FWS to withdraw proposed listings for the plants.

Changes made in that year also made the process of designating critical habitats more cumbersome. Since 1978, these habitats can be designated only after consideration of economic and other relevant impacts. Increased concern for procedural virtue is also evident in Congress's mandate, imposed in late 1988, that all revised or newly developed recovery plans be subject to public review and comment.

The process of listing species could be simplified, and the burden of proof placed on those who oppose listing rather than on the FWS or the NMFS. This change alone could benefit several thousand candidate species that the FWS has identified as being in need of listing. Giving the FWS a freer hand in listing would undoubtedly mean that some species might be classified as endangered even though they might not deserve to be. Consider the alternative: not listing many species that should be listed is a far worse choice when failure to act could easily mean extinction, an irreversible and potentially catastrophic mistake. Delisting a species inappropriately identified as endangered would be easily corrected and far less damaging.

Third, rather than requiring only federal agencies and their permit applicants to conserve and protect listed species through Section 7's consultation process, this requirement could be extended to all projects, public and private, that affect vulnerable species or their habitats. The present statutory arrangements create a potentially tragic irony. The Endangered Species Act states that species deserve to survive, but the same law then obligates only federal agencies to do anything to conserve listed species. If such species do deserve survival, then why neglect the nonfederal activities that just as readily lead to premature extinctions? Similarly, the consultation process could be strengthened to give the FWS a greater role in determining the fate of projects that lead to jeopardy opinions, as is suggested at the end of chapter 8. These changes in the consultation process would also address some of the problems inherent in the program's present impact model.

Fourth, the penalties for violating the law could be increased. Destruction of certain endangered plants became a crime only in late 1988. Until that time as well, the maximum penalty for *knowingly* violating the Endangered Species Act was only $10,000 per civil violation and $20,000 per criminal violation, amounts often far less than the potential profits that could be made from illegal activities.[10] The potential fines were increased by a factor of two and one-half in 1988. How frequently fines are imposed is a different matter. During the years of the Reagan presidency, both the number of federal enforcement agents and the number of investigations declined considerably because of alleged budgetary pressures.[11] Larger budgets could be used to hire more enforcement agents, and larger fines offer increased reason to pursue and punish illegal activities. Beginning in late 1988, all fines and penalties collected beyond the first $500,000 can be used for recovery efforts.

Fifth, statutory deadlines could be established for the completion of recovery plans, perhaps within two years of a species' listing. A subsequent deadline of a further year could be established for the full implementation of each plan. Just as the elderly are entitled to Social Security benefits, species with recovery plans would be entitled to benefit from the remedial measures necessary to enhance their prospects for survival. In addition, the FWS and NMFS could be given the authority to compel state and federal agencies to implement those portions of recovery plans subject to their control. When a state's cooperation is required, matching federal grants could be used as incentives.

Sixth, Congress could mandate a heightened visibility for efforts to protect endangered species within the Department of the Interior. Making some of the changes recommended above would contribute to this visibility. Congress could also require, however, that a separate service or bureau be created for the protection of biological diversity, thus reducing the number of officials to which the program is now subordinate.[12] Giving the new bureau or service complete administrative responsibility for all listing, consultation, and recovery activities would address the apparent reluctance and recalcitrance of the NMFS to implement the program. This change would have the further advantage of reducing the number of agencies involved in the drafting of regulations. As it is now, the FWS and NMFS must often issue joint regulations, such as on the consultation process. The heads of two separate agencies, often with disparate goals, must now approve these regulations, and the consequence is extended negotiation, reconciliation of competing policy agendas, and delay in issuing the regulations.

Additional changes could easily be recommended, but the ones mentioned provide a flavor of what is possible without jettisoning existing legal arrangements. Even if all the recommended changes are made in the near future, one would still have to ask if the changes could be implemented effectively and in sufficient time to stem the rising tide of premature extinctions in the United States. Changing a law does not lead to the strengthening of an agency's administrative backbone or its willingness to enforce a law aggressively. Large doses of unwavering political support from Congress and the president are vital to the resolution of these concerns.

More important, solutions to the problems of extinction are probably well beyond the marginal tinkering that now characterizes most policymaking. The Ehrlichs suggest that unless appropriate steps are taken soon to preserve species, then "humanity faces a catastrophe fully as serious as an all-out thermonuclear war."[13] Many will complain of hyperbole and exaggeration, but what if the Ehrlichs are correct? In such a case, solutions will be available only to the extent that policymakers abandon traditional approaches to environmental problems to address what Lettie Wenner has labeled the primary or societal causes of environmental degradation. For her, these are technology, population size, and economic growthmanship.[14] To deal with these concerns there is no shortage of recommendations, ranging from proposals to abolish the most cherished civil rights and liberties, to reduce the world's population far below current levels, to dependence on specially selected ecological mandarins with dictatorial powers to govern in an ecologically sound manner.[15] So radical are most of the proposed solutions that no government has yet given them serious consideration. This neglect displays elementary political expediency. As long as the pace of anthropogenic extinctions continues to accelerate, however, this neglect foolishly ignores ecological imperatives at a time when the number of effective choices is diminishing.

Perhaps human ingenuity and the earth's remarkable resilience can provide the means to escape what Robert Heilbroner calls "the descent into hell."[16] Whatever the reaction to accelerating rates of human-caused extinctions, responses are likely to establish precedents for such other environmental problems as the greenhouse effect, destruction of tropical rain forests, and ozone depletion due to the use of chlorofluorocarbons.[17] Together with anthropogenic extinctions, each of these global environmental problems offers the prospect of speculative but potentially cataclysmic ecological consequences at some uncertain date in the future. Proof of these consequences will not be available until they occur. This may mean that action in the present will be required to forestall events about which policymakers have limited information and familiarity. Furthermore, each problem is likely to require coordinated and far-ranging responses, not just from a few governments, but rather from many. Effective responses are undoubtedly vital. Seeking such responses will quite likely challenge the resolve and capacity of governmental institutions in many unpredictable and unprecedented ways.

Responses to extinction also provide answers to an unavoidable question: Are there obligations or responsibilities to future generations? Do those living today have a self-proclaimed right to deprive the yet unborn of their rightful inheritance, a healthy and habitable environment that freely provides all the necessities of life? As more species join the ranks of the extinct, answers become much less ambiguous. No other Americans have so rapidly or unconscionably depleted their biological endowment as has the present generation. Perhaps those born in the next ten, twenty, or thirty years will accept their diminished legacy without rancor or resentment, but it demands unrivaled arrogance and presumptuousness to expect them to do so. As they review what their parents and grandparents have done, they will realize that hundreds and perhaps even thousands of species, some with the potential to solve an infinite variety of human ills, were carelessly and unceremoniously relegated to oblivion. As former U.S. Senator

Gaylord Nelson once thoughtfully asked, "Is not a million years . . . of evolving landscape and fragile beauty worthy of our most attentive stewardship?"[18] He adds, "The ultimate test of man's conscience may be his willingness to sacrifice something today for future generations whose words of thanks will not be heard."

Appendix
Public Opinion and Preservation of Species

In 1982, Research and Forecasts, Inc., of New York City, conducted a nationwide survey of 1,310 Americans, age 18 and over, using randomly selected telephone numbers. The survey results are reported in *The Continental Group Report: Toward Responsible Growth* (Stamford, Conn.: The Continental Group, 1982). Due to sampling fluctuations, the responses were weighted to match the U.S. population on age, gender, ethnic origin, and education. The margin of error for the survey results is less than 3 percent. Professor Riley E. Dunlap of Washington State University, who was a member of the advisory panel for Research and Forecasts, generously provided access to the data reported in tables A.1 through A.3. Respondents were asked to agree or disagree with the following statement, which is used as the independent variable: "We must prevent any type of animal from becoming extinct, even if it means sacrificing some things for ourselves." Those people who did not know or who did not express an opinion are excluded from the analysis in all the tables that follow.

Table A.1   Attitudes toward Preserving Species and Modifying the Environment (in percentages)

	We must prevent extinction . . .	
	Agree/ Strongly agree	Disagree/ Strongly disagree
Modifying the environment for human uses seldom causes serious problems.		
Agree/Strongly agree	59.0	59.3
Disagree/Strongly disagree	41.0	40.7
Total	100.0	100.0
(*N*)	971	241

$\chi^2 = .0085$, not significant, Somer's d $= -.0032$

Table A.2   Attitudes toward Preserving Species and Community Growth (in percentages)

	We must prevent extinction . . .	
	Agree/ Strongly agree	Disagree/ Strongly disagree
Thinking about new housing and inviting new industry to your community, do you think that your community ought to continue to grow or should not grow?		
Continue to grow	60.1	64.8
Should not grow	39.9	35.2
Total	100.0	100.0
(*N*)	1,017	247

$\chi^2 = 1.844$, not significant, Somer's d $= -.047$

Table A.3   Attitudes toward Preserving Species and Humankind's
Domination of Nature (in percentages)

	We must prevent extinction . . .	
	Agree/ Strongly agree	Disagree/ Strongly disagree
Mankind was created to rule over the rest of nature.		
Agree/Strongly agree	54.8	62.5
Disagree/Strongly disagree	45.2	37.5
Total	100.0	100.0
(N)	1,014	248

$\chi^2 = 4.76$, p $= .05$, Somer's d $= -.077$

Table A.4   Trade-Offs Between the Protection of Endangered Fish and
Various Uses of Water (in percentages)

	Water Use				
	Dammed to make a lake for recreational use	Diverted to cool industrial plant machinery	Dammed to provide hydro-electric power	Diverted to irrigate agricultural crops	Diverted to increase human drinking supplies
Approve[a]	40.6	50.8	75.3	85.7	89.0
Disapprove[b]	59.4	49.2	24.7	14.3	11.0
Total	100.0	100.0	100.0	100.0	100.0
(N)	2,374	2,316	2,336	2,384	2,391

Source: Stephen R. Kellert, "Public Attitudes Towards Critical Wildlife and National Habitat Issues," in Senate Committee on Environment and Public Works, hearings, *Endangered Species Act Oversight*, 97th Cong., 1st sess., 1981, 390.

Note: Respondents were presented with the following statement: "Various kinds of fish have been threatened with extinction because of dams, canals, and other water projects. Please indicate if you would approve of the following water uses if they were to endanger a species of fish."

[a] Includes those who strongly approved, approved, or slightly approved.

[b] Includes those who strongly disapproved, disapproved, or slightly disapproved.

With the exception of water use for industrial machinery, all the differences are statistically significant with p $< .0001$.

Table A.5   Willingness to Consider Trade-Offs in Protection of
Endangered Species (in percentages)

Attitude toward Protection	Bald eagle	Agassiz trout	American crocodile	Eastern indigo snake	Furbish lousewort	Kauai wolf spider
			Species			
Favor	91.7	77.6	73.6	47.2	41.6	37.5
Oppose	8.3	22.4	26.4	52.8	58.4	62.5
Total	100.0	100.0	100.0	100.0	100.0	100.0
(N)	2,382	2,249	2,325	2,255	2,000	2,226

Source: Same as table A.4, 382–83.
Note: Respondents were asked to respond to the following statement: "A recent law passed to protect endangered species may result in changing some energy development projects at greater cost. As a result, it has been suggested that endangered species protection be limited only to certain animals and plants. Which of the following endangered species would you favor protecting, even if it resulted in higher costs for an energy development project?"

Table B.1   Average Length of Time in Days to Designate Critical Habitats

Taxa	Pre–1978 Act Average	Standard Deviation	(N)	Post–1978 Act Average	Standard Deviation	(N)
Mammals	296.75	33.24	4	483.13	360.18	8
Birds	384.11	242.62	9	983.00	0	1
Reptiles	232.71	60.65	7	527.67	428.49	6
Amphibians	230.00	17.32	3	117.00	0	1
Fishes	262.63	132.42	8	509.87	273.22	30
Crustaceans	NA	NA	0	1,090.00	0	1
Insects	NA	NA	0	321.80	310.71	5
Plants	569.00	0	2	583.96	494.11	23
Totals	309.15	166.10	33	527.43	381.16	75

Note: Includes all critical habitats designated through 1989, except the chinook salmon, for which the NMFS established a critical habitat on an immediate, but temporary emergency basis in 1989.

Table C.1   Mean Interval in Years between Listing Date and
Approval of Recovery Plan

Year(s) Plan Approved	Mean Years	(N)
1975–78	9.58	19
1979	9.96	11
1980	9.26	10
1981	10.00	5
1982	9.06	27
1983	9.74	42
1984	8.81	79
1985	6.81	28
1986	6.23	19
1987	2.74	17
1988	2.51	26
1989	2.38	6
Total	7.68	289

Note: Includes all recovery plans approved through 1989, except for delisted species.
Minimum time: 9.5 months (smoky madtom)
Maximum time: 18.5 years (San Francisco garter snake)
Median and mode: 6.5 years

Table C.2   Listing Year by Year of Recovery Plan Approval

Approval Year(s)	Listing Year(s)				
	1967	1970	1973–78	1979–80	1981–89
1975–78	14	5	—	—	—
1979–80	11	5	5	—	—
1981	2	1	1	1	—
1982	8	4	12	3	—
1983	14	4	19	5	—
1984	11	10	42	14	2
1985	2	—	11	9	6
1986	—	3	2	5	9
1987	—	—	—	2	15
1988	—	—	—	—	26
1989	—	—	—	—	6
Totals	62	32	92	39	64

Note: Includes all recovery plans approved through 1989.

Notes

1 Biological Diversity as an Ecological Issue

1 Estimating the total number of species and subspecies is difficult. Estimates range from a few million to well over thirty million. For the latter estimate, see Terry L. Erwin, "Tropical Forest Canopies: The Last Biotic Frontier," *Bulletin of the Entomological Society of America* 29 (1983): 14–19. Of the total number of species and subspecies, less than two million have been formally named and taxonomically described. Australia, as an example, is believed to have about 108,000 insect species, but less than half have been named. Nearly 40 percent remain unnamed and even uncollected. Cited in Edward O. Wilson, "The Biological Diversity Crisis," *BioScience* 35 (Dec. 1985): 701. In South America, the estimated number of undiscovered species of insects ranges from five to fifty million, according to Michael A. Mares, "Conservation in South America: Problems, Consequences, and Solutions," *Science* 233 (15 Aug. 1986): 737.

2 *FR* 47 (15 Jan. 1982): 2317–18.

3 *ESTB* 3 (Aug. 1978): 8.

4 David M. Raup, "Biological Extinction in Earth History," *Science* 231 (28 Mar. 1986): 1528–33. The normal or background rate of extinction is the subject of disagreement. In contrast to Raup's findings, Edward C. Wolf claims that the background rate is only a "few species per million years." See Wolf, *On the Brink of Extinction: Conserving the Diversity of Life* (Washington: Worldwatch Institute, 1987), 6.

5 Norman Myers, *The Sinking Ark* (Oxford: Pergamon, 1979), 4; Paul Ehrlich and Anne Ehrlich, *Extinction: The Causes and Consequences of the Disappearance of Species* (New York: Random House, 1981), 26.

6 Myers, *Sinking Ark*, 4.

7 Ibid.

8 Wilson, "Biological Diversity Crisis," 703; Thomas Lovejoy, "We Must Decide Which Species Will Go Forever," *Smithsonian* 7 (July 1976): 55.

9 For estimates of the number of anthropogenic extinctions, see Erik Eckholm, "The Age of Extinction," *The Futurist* 12 (Oct. 1978): 289 (500,000 species to be

lost by the end of this century); Margery Oldfield, *The Value of Conserving Genetic Resources* (Washington: NPS, 1984), 285 (500,000 to 1,000,000 species by the beginning of the next century); Norman Myers, "The Exhausted Earth," *Foreign Policy* 42 (Spring 1981): 141 (1,000,000 species by 2000); Thomas Eisner, cited in Peggy Eastman and Tineke Bodde, "Endangered Species Act," *BioScience* 32 (Apr. 1982): 246 (1,000,000 species by the end of this century); and Donald R. Davis, cited in Myers, *Sinking Ark*, 20 (about three million insect species by 2000). For estimates of daily extinction rates, see Myers, *Sinking Ark*, 5, 31; and Wolf, *Brink of Extinction*, 6 ("several hundred extinctions per day in the next 20 to 30 years"). Whatever the estimated number of human-caused extinctions, the numbers are subject to intense debate and disagreement. The OTA notes that "discrepancies among the estimated extinction rates have called into question the credibility of all such estimates." See OTA, *Technologies to Maintain Biological Diversity* (Washington, 1987), 74. For a discussion of these estimates, see Michael Harwood, "Math of Extinction," *Audubon* 84 (Nov. 1982): 18–21; Mares, "Conservation in South America," 734–39; and Julian L. Simon, "Natural Resources and Population: What Are the Proper Roles of Public and Private Sectors?" in M. N. Maxey and R. L. Kuhn, eds., *Regulatory Reform: New Vision or Old Curse?* (New York: Praeger, 1985), 77–79.

10   See Harold Koopowitz and Hilary Kaye, *Plant Extinction: A Global Crisis* (Washington: Stone Wall Press, 1983), 6 (anticipated loss of between 15 and 25 percent of all higher plant species by 2000); Laura Tangley, "A New Plan to Conserve Earth's Biota," *BioScience* 35 (June 1985): 334 ("at least 20 % of all today's species could be gone within 20 years"). Norman Myers, who is among the most prolific and pessimistic estimators, believes that "we can anticipate witnessing the disappearance of probably one-quarter, possibly one-third, and conceivably a still larger share of all extant species." Myers, "Extinction Rates Past and Present," *BioScience* 39 (Jan. 1989): 39. Elsewhere Myers emphasizes that "it is possible that we will lose one-third, and conceivable that we will lose one-half of all species." "Genetic Resources in Jeopardy," *Ambio* 13 (1984): 174.

11   Lee M. Talbot, cited in S. Dillon Ripley, "Extinction's Tide and the Ripples and Eddies of Hope," *Smithsonian* 2 (Feb. 1972): 24. At the time of Talbot's statement, he was an ecologist at the Smithsonian Institution and a senior scientist on the CEQ.

12   CEQ, *The Global 2000 Report to the President*, Vol. 2, *The Technical Report* (Washington, 1980), 331. The Department of State accepted this figure when it sponsored a conference on biological diversity in 1981. See the statement of James L. Buckley, under secretary of state for security assistance, science, and technology, in Department of State, *Proceedings of the U.S. Strategy Conference on Biological Diversity* (Washington, 1982), 9.

13   OTA, *Biological Diversity*, 3.

14   World Commission on Environment and Development, *Our Common Future* (New York: Oxford Univ. Press, 1987), 148.

15   House Committee on Merchant Marine and Fisheries, hearings, *Predatory Mammals and Endangered Species*, 92nd Cong., 2nd sess., 1972, 572.

16   House Committee on Merchant Marine and Fisheries, *Endangered and Threatened Species Conservation Act of 1973*, 93rd Cong., 1st sess., 1973, H. Rept. 93–412, 2;

and House Committee on Merchant Marine and Fisheries, *Endangered Species Act Amendments of 1978*, 95th Cong., 2nd sess., 1978, H. Rept. 95–1625, 5.

17 Letters from Elaine Joyal, The Nature Conservancy Science Department, to the author, 29 July 1988 and 9 Sept. 1988. See also Philip Shabecoff, "Survey Finds Native Plants in Imminent Peril," *New York Times*, 6 Dec. 1988, C1.

18 *ESTB* 10 (Nov. 1985): 1; Koopowitz and Kaye, *Plant Extinction*, 7; and Andrew J. Berger, "Hawaii's Dubious Distinction," *Defenders* 50 (Dec. 1975): 495.

19 CEQ, *Environmental Quality, 1985*, (Washington, 1986), 298. The OTA reached essentially the same conclusions in *Biological Diversity*, 294–96.

20 Myers, *Sinking Ark*, 158. Myers estimated that U.S. consumption of tropical hardwoods would be approximately fifteen million cubic meters in 1990, almost nineteen times as much as was imported in 1950.

21 R. Aiken and M. Moss, "Man's Impact on the Tropical Rainforests of Peninsular Malaysia: A Review," *Biological Conservation* 8 (1975): 213–29.

22 U.S. imports of zinc between 1983 and 1987 totaled almost four million short tons. The crustal abundance of zinc is only 0.0082 percent by weight, meaning that 12,195 tons of rock must be processed to produce one ton of zinc, assuming perfect efficiency in the recovery process. See William Ophuls, *Ecology and the Politics of Scarcity* (San Francisco: W. H. Freeman, 1977), 69–70.

23 Ehrlich and Ehrlich, *Extinction*, 194–98; and Ginette Hemley, "International Wildlife Trade," in William J. Chandler, ed., *Audubon Wildlife Report, 1988/89* (San Diego: Academic Press, 1988), 337–74.

24 Oldfield, *Genetic Resources*, 280.

25 *ESTB* 6 (Apr. 1981): 6.

26 *ESTB* 7 (June 1982): 1, 3–4.

27 John Noble Wilford, "Tiny Fish Looms Large in Evolution Debate," *New York Times*, 20 Apr. 1982, C1–2.

28 Myers, *Sinking Ark*, 4; Ehrlich and Ehrlich, *Extinction*, 7.

29 Wilson, "Biological Diversity Crisis," 701.

30 Edward O. Wilson, *Biophilia*, (Cambridge: Harvard Univ. Press, 1984), 130; Ehrlich and Ehrlich, *Extinction*, 5.

31 See Myers, *Sinking Ark*, ch. 4 and 5; Ehrlich and Ehrlich, *Extinction*, ch. 3 and 4; OTA, *Biological Diversity*, ch. 2; and Robert Prescott-Allen and Christine Prescott-Allen, *What's Wildlife Worth?* (London: International Institute for Environment and Development, 1982). Much of what follows, which is not intended to be exhaustive, relies on these sources.

32 See Peter H. Raven, director, Missouri Botanical Garden, in Senate Committee on Environment and Public Works, hearings, *Endangered Species Act Oversight*, 97th Cong., 1st sess., 1982, 293; Erwin, "Tropical Forest Canopies."

33 Ehrlich and Ehrlich, *Extinction*, 94.

34 Norman R. Farnsworth and D. D. Soejarto, "Potential Consequences of Plant Extinction in the United States on the Current and Future Availability of Prescription Drugs," *Economic Botany* 39 (1985): 231–40; Norman Myers, *A Wealth of Species* (Boulder: Westview, 1983).

35 Myers, *Sinking Ark*, 69.

36 It is humanistic to assume that species should have economic or utilitarian value in order to justify their continued existence. This theme and a contrasting view

are discussed in ch. 4, and in David W. Ehrenfeld, *The Arrogance of Humanism* (New York: Oxford Univ. Press, 1978), 177–79.

37 Wilson, "Biological Diversity Crisis," 702.

38 National Academy of Sciences, *Underexploited Tropical Plants of Promising Economic Value* (Washington, 1975); Norman Myers, ed., *GAIA: An Atlas for Planet Management* (Garden City, N.J.: Doubleday, 1984).

39 Myers, *Sinking Ark*, 6.

40 FWS and Bureau of the Census, *1980 National Survey of Fishing, Hunting, and Wildlife-Associated Recreation* (Washington, 1982), 31, 106.

41 Edward O. Wilson, cited in Ehrlich and Ehrlich, *Extinction*, 3. Robert Allen concurs. He believes the "devastation of the biosphere is ultimately the greatest of all threats to the survival and well-being of human beings." *How to Save the World: Strategy for World Conservation* (Totowa, N.J.: Barnes & Noble, 1980), 15.

42 David W. Ehrenfeld, *Conserving Life on Earth* (New York: Oxford Univ. Press, 1972), 158.

2 Biological Diversity as a Political Issue

1 Christopher D. Stone, *Where the Law Ends: The Social Control of Corporate Behavior* (New York: Harper & Row, 1975), xii. Italics omitted. See also John Passmore, *Man's Responsibility to Nature: Ecological Problems and Western Traditions* (New York: C. Scribner's Sons, 1974), 53, 57.

2 Garrett Hardin, "The Tragedy of the Commons," *Science* 162 (13 Dec. 1968): 1243–48.

3 Oran Young, *Natural Resources and the State: The Political Economy of Resource Management* (Berkeley and Los Angeles: Univ. of California Press, 1981), 45. See also Rice Odell, "Can Science Deal with Environmental Uncertainties?" *Conservation Foundation Letter* (Jan. 1982): 2–3.

4 John W. Kingdon, *Agendas, Alternatives, and Public Policies* (Boston: Little, Brown, 1984), 3.

5 Ibid., 69. See also Paul Light, *The President's Agenda* (Baltimore: Johns Hopkins Univ. Press, 1981), 11.

6 Roger W. Cobb and Charles D. Elder, *Participation in American Politics: The Dynamics of Agenda Building*, 2nd ed. (Baltimore: Johns Hopkins Univ. Press, 1983), 177. See also Kingdon, *Agendas*, 177.

7 James Q. Wilson, "The Politics of Regulation," in J. W. McKie, ed., *Social Responsibility and the Business Predicament* (Washington: Brookings Institution, 1974), 146–47.

8 Paul Ehrlich and Anne Ehrlich, *Extinction: The Causes and Consequences of the Disappearance of Species* (New York: Random House, 1981), 242.

9 Kingdon, *Agendas*, 52; Cobb and Elder, *Participation*, 11.

10 Wilson, "Politics of Regulation," 140. See also Mancur Olson, Jr., *The Logic of Collective Action* (New York: Schocken, 1968), 21.

11 Kingdon, *Agendas*, 100.

12 Keith Hawkins, *Environment and Enforcement: Regulation and the Social Definition of Pollution* (Oxford: Clarendon, 1984), 12.

13 Kingdon, *Agendas*, 179. In the words of William T. Conway, director of the New

York Zoological Park, "man can get along economically without the overwhelming majority of animals and plant life for some period of time." In *Conservation Foundation Letter* (May 1978): 7.

14   In *Extinction*, Ehrlich and Ehrlich acknowledge that one of "the great problems of ecologists today is their inability in most cases to predict the consequences to an ecosystem of the extinction of any given population or species. . . . What limited knowledge is available indicates that the consequences of the extermination of any given group of organisms are . . . likely to be unique" (96). See also Department of State, *Proceedings of the U.S. Strategy Conference on Biological Diversity* (Washington, 1982), 9.

15   Ehrlich and Ehrlich, *Extinction*, 179. For a discussion of ecological illiteracy, see N. W. Moore, "Experience with Pesticides and the Theory of Conservation," *Biological Conservation* 1 (1969): 202–3; and Garrett Hardin, *The Limits of Altruism* (Bloomington: Indiana Univ. Press, 1977), 34.

16   Wilson, "Politics of Regulation," 139.

17   "Old, But Far From Feeble," *The Economist* 306 (12 Mar. 1988): 30.

18   Daniel A. Mazmanian and Paul A. Sabatier, *Implementation and Public Policy* (Glenview, Ill.: Scott, Foresman, 1983), 24.

19   Norman Myers, *The Sinking Ark* (Oxford: Pergamon, 1979), 233.

20   Lettie McS. Wenner, *One Environment Under Law* (Pacific Palisades, Calif.: Goodyear, 1976), 114.

21   Anthony Downs, *Inside Bureaucracy* (Boston: Little, Brown, 1966), 206; Cobb and Elder, *Participation*, 164.

22   Ehrlich and Ehrlich, *Extinction*, 243. The Ehrlichs' opinion is widely shared. Edward C. Wolf believes that: "Embarking on a path away from mass extinction will require a radical departure from deeply embedded policies and land-use practices." See Wolf, *On the Brink of Extinction: Conserving the Diversity of Life* (Washington: Worldwatch Institute, 1987), 38; and Erik Eckholm, *Disappearing Species: The Social Challenge* (Washington: Worldwatch Institute, 1978), 11.

23   Wilson, "Politics of Regulation," 139–46; Downs, *Inside Bureaucracy*, 46.

24   Roy Gregory, *The Price of Amenity* (London: Macmillan, 1971), 302.

25   Helen Ingram and Dean Mann, "Political Responses to Environmental Scarcity," in S. Welch and R. Miewald, eds., *Scarce Natural Resources: The Challenge to Public Policymaking* (Beverly Hills, Calif.: Sage, 1983), 133. See also David W. Ehrenfeld, *The Arrogance of Humanism* (New York: Oxford Univ. Press, 1978), 17.

26   Susan Welch and Robert Miewald, "Natural Resources Scarcity: An Introduction," in Welch and Miewald, *Scarce Natural Resources*, 17–18.

27   See Julian L. Simon, *The Ultimate Resource* (Princeton, N.J.: Princeton Univ. Press, 1981), and Harold J. Barnett and Chandler Morse, *Scarcity and Growth* (Baltimore: Johns Hopkins Univ. Press, 1963). In *Arrogance of Humanism*, David Ehrenfeld discusses the humanist view that all finite resources have substitutes.

28   OTA, *Technologies to Maintain Biological Diversity* (Washington, 1987), 5. This report provides a comprehensive discussion of existing technologies. See also David Ehrenfeld, *Conserving Life on Earth* (New York: Oxford Univ. Press, 1972), 4.

29   Department of State, *Conference on Biological Diversity*, 105.

30   Carol H. Weiss, *Evaluation Research* (Englewood Cliffs, N.J.: Prentice-Hall, 1972), 2, 25. It should be noted that there is no single endangered species program.

Many federal agencies have at least some responsibility for protecting such species, but the FWS and NMFS are the most significant of these agencies.

31 Thomas R. Dye, "Policy Analysis and Political Science: Some Problems at the Interface," in S. S. Nagel, ed., *Policy Studies and the Social Sciences* (Lexington, Mass.: Lexington Books, 1975), 282–83.

32 Endangered Species Preservation Act of 1966, P.L. 89–669, Sections 1(a) and 1(b).

33 Endangered Species Act of 1973, P.L. 93–205, Sections 2(b) and 3(2).

34 Downs, *Inside Bureaucracy*; Graham Allison, *The Essence of Decision* (Boston: Little, Brown, 1971); and Michael Lipsky, "Standing the Study of Public Policy Implementation on Its Head," in W. D. Burnham and M. Weinberg, eds., *American Politics and Public Policy* (Cambridge: MIT Press, 1978).

35 Weiss, *Evaluation Research*, 48. See also David Nachmias, *Public Policy Evaluation* (New York: St. Martin's, 1979). 9–11.

36 Helen Ingram, "Implementation: A Review and Suggested Framework," in A. Wildavsky and N. B. Lynn, eds., *Public Administration: The State of the Field* (forthcoming).

37 Nachmias, *Public Policy Evaluation*, 18.

38 Edwin C. Hargrove, *The Missing Link: The Study of the Implementation of Social Policy* (Washington: Urban Institute, 1975), 114.

39 A native species is defined as: a) one that spends all or part of its life on land or in waters subject to the jurisdiction of the United States; or, b) a plant that is found solely in the United States or in the United States and at least one other country. Geographically, this includes the fifty states, Guam, Puerto Rico, the District of Columbia, and the Virgin and Mariana Islands.

40 Helen Ingram and Dean Mann, "Policy Failure: An Issue Deserving Analysis," in Ingram and Mann, eds., *Why Policies Succeed or Fail* (Beverly Hills, Calif.: Sage, 1980), 12.

41 Roger L. Di Silvestro, "Our Looming Failure on Endangered Species," *Defenders* 59 (July-Aug. 1984): 21. See also Michael Bean, Faith Campbell, and Robert Davison, "The Endangered Species Program Needs Rejuvenation," *Conservation Foundation Letter* (Jan.-Feb. 1985): 1.

42 Eugene Bardach, *The Implementation Game* (Cambridge: MIT Press, 1977), 251.

43 Letter from John J. Fay, staff botanist, Division of Endangered Species and Habitat Conservation, FWS, to the author, 1 Dec. 1987.

44 Department of State, *Conference on Biological Diversity*, 5.

45 Michael J. Bean, Environmental Defense Fund, in House Committee on Merchant Marine and Fisheries, hearings, *Endangered Species Act*, 99th Cong., 1st sess., 1985.

46 See also Myers, *Sinking Ark*, 232.

3 Administrative Capacity for Action

1 Francis E. Rourke, *Bureaucracy, Politics, and Public Policy*, 3rd ed. (Boston: Little, Brown, 1984), 91.

2 House Committee on Merchant Marine and Fisheries, hearings, *Reorganization of Fish and Wildlife Service*, 85th Cong., 1st sess.; 1957, 148.

3   Jeanne N. Clarke and Daniel McCool, *Staking Out the Terrain: Power Differentials Among Natural Resource Management Agencies* (Albany: State Univ. of New York Press, 1985), 83.

4   Ibid., 7.

5   Rourke, *Bureaucracy*, 99.

6   Department of the Interior, *The Interior Budget in Brief: Fiscal Year 1985 Highlights* (Washington, 1984), 1.

7   William K. Wyant, *Westward in Eden: The Public Lands and the Conservation Movement* (Berkeley and Los Angeles: Univ. of California Press, 1982), 102.

8   Nathaniel P. Reed and Dennis Drabelle, *The United States Fish and Wildlife Service* (Boulder, Colo.: Westview, 1984), 11.

9   Department of the Interior, *A Year of Progress: Preparing for the 21st Century* (Washington, 1983), 1.

10   Ibid.

11   Wyant, *Westward in Eden*, 57–58.

12   See, for example, Clarke and McCool, *Staking Out the Terrain*, 111–14; Phillip Foss, *Politics and Grass* (Seattle: Univ. of Washington Press, 1960); Grant McConnell, *Private Power and American Democracy* (New York: Knopf, 1966); and Paul J. Culhane and H. Paul Friesema, "Land Use Planning for the Public Lands," *Natural Resources Journal* 19 (Jan. 1979), 43–74.

13   BLM, *Managing the Public Lands* (Washington, 1980), 4.

14   Noel Rosetta, "Herds, Herds on the Range," *Sierra* 70 (Mar.-Apr. 1985): 44. See also McConnell, *Private Power*, 205.

15   BLM, *Range Condition Report Prepared for the Senate Committee on Appropriations* (Washington, 1975). A more recent report suggests that the situation remained largely unchanged in the late 1980s. See GAO, *Rangeland Management: More Emphasis Needed on Declining and Overstocked Grazing Allotments* (Washington, 1988).

16   GAO, *National Direction Required for Effective Management of America's Fish and Wildlife* (Washington, 1981), 47.

17   OTA, *Environmental Protection in the Federal Coal Leasing Program* (Washington, 1984). Also see two reports by the GAO, *Improvements Needed in Administration of Federal Coal-Leasing Program* (Washington, 1972) and *Further Action Needed on Recommendations for Improving the Administration of Federal Coal-Leasing Program* (Washington, 1975).

18   William E. Warne, *The Bureau of Reclamation* (New York: Praeger, 1973), 3.

19   Kenneth J. Meier, *Politics and the Bureaucracy*, 2nd ed. (Monterey, Calif.: Brooks/Cole, 1986), 100. As a measure of clientele support, Meier calculated the number of interest groups that testified in support of over one hundred federal agencies at appropriations hearings in the House between 1974 and 1976. Among all agencies ranked, the BOR was second.

20   Clarke and McCool, *Staking Out the Terrain*, 137–42.

21   Reed and Drabelle, *Fish and Wildlife Service*, 12.

22   Senate Committee on Interior and Insular Affairs and Committee on Commerce, hearings, *Interior Nomination*, 93rd Cong., 2nd sess., 1974, 10.

23   The Animal Damage Control Program was transferred to the Department of Agriculture in 1986, partly in response to demands from farmers, woolgrowers, and cattlemen. For a discussion of the program, see Ruth Norris, "The Federal

Animal Damage Control Program," in Roger L. Di Silvestro, ed., *Audubon Wildlife Report* (New York: Academic Press, 1987), 223–37.

24 For a discussion of the 1080 controversy, see Mark Palmer, "Compound 1080 —Poison Returns to the Range," *Sierra* 67 (Nov.-Dec. 1982): 25–30; and "Watt and Gorsuch Go After Coyotes," *Not Man Apart* 12 (Jan. 1982): 4. The poison is usually put in a dead lamb, which serves as bait for coyotes.

25 GAO, *National Wildlife Refuges: Continuing Problems with Incompatible Uses Call for Bold Action* (Washington, 1989).

26 Ibid., 26.

27 GAO, *Economic Uses of the National Wildlife Refuge System Unlikely to Increase Significantly* (Washington, 1984), 58–59.

28 Dale Russakoff, "Under Siege: Threats to Refuges Enumerated," *Washington Post*, 8 Aug. 1983, A11; Philip Shabecoff, "Toxic Water Threatens Many Wildlife Refuges," *New York Times*, 30 Apr. 1985, A16; and Shabecoff, "Survey Finds 10 Wildlife Refuges Contaminated by Toxic Materials," *New York Times*, 4 Feb. 1986, A15. See also GAO, *Wildlife Management: National Refuge Contamination is Difficult to Confirm and Clean Up* (Washington, 1987).

29 The problems at Kesterson are discussed in Robert Lindsey, "Water That Enriched Valley Becomes a Peril in California," *New York Times*, 7 Jan. 1985, A1; and Lindsey, "Irrigation Water Cut Off by U.S. to Protect Birds," *New York Times*, 16 Mar. 1985, 7. See also Keith Schneider, "Crisis at Kesterson," *The Amicus Journal* 7 (Fall 1985): 22–27.

30 Cited in Stephen Fox, *John Muir and His Legacy: The American Conservation Movement* (Boston: Little, Brown, 1981), 283.

31 House Committee on Appropriations, hearings, *Energy and Water Development Appropriations for 1982*, 97th Cong., 1st sess., 1981, 7–9.

32 GAO, *Economic Uses of the Refuge System*, 8. This report discusses in detail FWS efforts to expand use of the refuges. For discussions of the public and commercial demands placed on the refuges, see Dennis Drabelle, "The National Wildlife Refuge System," in Roger L. Di Silvestro, ed., *Audubon Wildlife Report 1985* (New York: National Audubon Society, 1985), 161–66.

33 Eric Lichtblau, "Federal Report Blasts Offshore Oil Studies," *Los Angeles Times*, 4 June 1988, sec. 1, 32; Alan L. Miller, "Hodel Takes Aim at Endangered Species Act," *Los Angeles Times*, 4 Feb. 1988, sec. 1, 3; Martin Tolchin, "Manuel Lujan, Jr.," *New York Times*, 23 Dec. 1988, A25.

34 Meier, *Politics and the Bureaucracy*, 55–56.

35 Rourke, *Bureaucracy*, 99.

36 CEQ et al., *Public Opinion on Environmental Issues: Results of a National Opinion Survey* (Washington, 1980), 18.

37 *The Continental Group Report, Toward Responsible Growth: Economic and Environmental Concern in the Balance* (Stamford, Conn.: Continental Group, 1982), 19. The survey had 1,310 respondents.

38 Stephen R. Kellert, "Social and Perceptual Factors in the Preservation of Animal Species," in Bryan G. Norton, ed., *The Preservation of Species: The Value of Biological Diversity* (Princeton, N.J.: Princeton Univ. Press, 1986). See also Kellert's "Americans' Attitudes and Knowledge of Animals," in *Transactions of the*

45th North American Wildlife and Natural Resources Conference (Washington: Wildlife Management Institute, 1980), 111–24.

39 Kellert, "Social and Perceptual Factors," 57. Italics in original.

40 Daniel A. Mazmanian and Paul A. Sabatier believe that "legislative policy committees become increasingly sympathetic to target groups over time, in part as a reflection of changes in the balance of interest group support and in part because constituency casework appears to be weighted toward complaints." See *Implementation and Public Policy* (Glenview, Ill.: Scott, Foresman, 1983), 34.

41 Keith Schreiner, "Looking Back over My Shoulder," ESTB 4 (May 1979): 5.

42 Rourke, *Bureaucracy*, 103.

43 John D. Leshy, "Natural Resource Policy," in Paul R. Portney, ed., *Natural Resources and the Environment: The Reagan Approach* (Washington: Urban Institute, 1984), 19.

44 Jim Doherty, "Refuges on the Rocks," *Audubon* 85 (July 1983): 96.

45 GAO, *National Direction Required*, p. 25.

46 Senate Committee on Environment and Public Works, hearings, *Nomination of Frank H. Dunkle*, 99th Cong., 2nd sess., 1986, 5–7. In fairness, it must be mentioned that there are examples to the contrary. When Robert A. Jantzen was nominated to be director of the FWS in 1981, he received endorsements from the Environmental Defense Fund, the National Wildlife Federation, and the National Audubon Society during his confirmation hearings. George Bush's choice to head the FWS, John Turner, likewise received support from several environmental groups.

47 Fox, *John Muir and His Legacy*, 262. Fox also notes that the National Audubon Society once agreed to accept an annual subvention of $25,000 from the Winchester Repeating Arms Company in exchange for the society's agreement to increase its efforts to protect game birds. After the agreement was publicized, controversy ensued, and the Audubon Society decided not to accept the money. See Fox, 155–56.

48 Defenders of Wildlife, *Saving Endangered Species: A Report and Plan for Action* (Washington: Defenders of Wildlife, 1984); Doherty, "Refuges on the Rocks," 74–116.

49 Richard Fenno, *The Power of the Purse* (Boston: Little, Brown, 1966), 363. Fenno also noted: "There would seem to be no question that [the Interior] department was markedly less successful than any of the other 6 studied in terms of receiving what it requested from the House [Appropriations] Committee. And this departmental weakness reflects itself across the board in terms of its various subunits" (369).

50 Clarke and McCool, *Staking Out the Terrain*, 128, 132.

51 Office of the Federal Register, *Weekly Compilation of Presidential Documents* 25 (19 June 1989): 892–93.

52 Senate Committee on Appropriations, hearings, *Department of the Interior and Related Agencies Appropriations for 1977, Part 3*, 94th Cong. 2nd sess., 1976, 307.

53 Rourke, *Bureaucracy*, 111–12.

54 House Committee on Appropriations, hearings, *Energy and Water Development Appropriations for 1981, Part 3*, 97th Cong., 1st sess., 1981, 3431.

55 Ibid., 3462.

56 Ibid., 8.

57 Ibid., 3439.

58 Clarke and McCool, *Staking Out the Terrain*, 139.

59 On this topic generally, see Rourke, *Bureaucracy*, 94–95.

60 Roger W. Cobb and Charles D. Elder, *Participation in American Politics: The Dynamics of Agenda Building*, 2nd ed. (Baltimore: Johns Hopkins Univ. Press, 1983), 99.

61 Much of what follows is from Dean Schooler, Jr., *Science, Scientists, and Public Policy* (New York: Free Press, 1971).

62 Ibid., 44. Italics omitted.

63 GAO, *National Direction Required*, 10.

64 Ibid., 11. Shortage of staff and scientific expertise have long been problems. Richard A. Cooley noted that in the 1930s the Bureau of Fisheries (then in the Department of Commerce) had such limited funds "that management of the [Alaskan] salmon fisheries had been largely a matter of trial and error." Cooley added that in the 1950s, "the funds expended by industry for biological research exceeded those spent by the federal government. This resulted in a situation in which the regulating agency [the FWS], in the process of setting policy, frequently was obliged to rely upon the scientific findings produced by those who were being regulated." See Cooley, *Politics and Conservation: The Decline of the Alaska Salmon* (New York: Harper & Row, 1963), 153, 163.

65 GAO, *National Direction Required*, 14. In 1983, in response to a congressional inquiry, the FWS reported that other agencies fully accepted its recommendations on permits and licenses about 20 percent of the time, partially in about 50 percent of the applications, and in a less than a satisfactory manner in the remaining instances. See House Committee on Appropriations, hearings, *Department of the Interior and Related Agencies Appropriations for 1984, Part 6*, 98th Cong., 1st sess., 1983, 578.

66 GAO, *National Direction Required*, 16–17.

67 Norman D. Levine, "Evolution and Extinction," *BioScience* 39 (Jan. 1989): 38.

68 Schooler, *Science, Scientists*, 51.

69 Julian L. Simon and Aaron Wildavsky, "On Species Loss, the Absence of Data, and Risks to Humanity," in Julian L. Simon and Herman Kahn, eds., *The Resourceful Earth* (New York: Basil Blackwell, 1984), 176.

70 Ibid., 172. For an appraisal of Simon's views on endangered species, see Thomas E. Lovejoy, "Species Leave the Ark One by One," in Norton, *The Preservation of Species*, 14.

71 Graham T. Allison, *Essence of Decision: Explaining the Cuban Missile Crisis* (Boston: Little, Brown, 1971), 72.

72 Ibid., 93–94. Other authors who make a similar point include Mazmanian and Sabatier, *Implementation*, 27, and Downs, *Inside Bureaucracy*, 212–20.

73 Letter from Jack P. Woolstenhulme, acting regional director, FWS Region 2, Albuquerque, to Joe Winkle, FDAA, Dallas, 28 Feb. 1979.

74 Eugene Bardach and Robert A. Kagan, *Going by the Book: The Problems of Regulatory Unreasonableness* (Philadelphia: Temple Univ. Press, 1982), 42.

75 Ibid., 42–43.

76   Peter H. Rossi, Howard E. Freeman, and Sonia R. Wright, *Evaluation: A Systematic Approach* (Beverly Hills, Calif.: Sage, 1979), 115.

4   Matching Resources and Priorities

1   Paul Ehrlich and Anne Ehrlich, *Extinction: The Causes and Consequences of the Disappearance of Species* (New York: Random House, 1981), 243–44.

2   The intent here is not to dismiss casually these possibilities. It is simply the case that many of these changes are not likely to occur in time to save many species. For indications of future changes, see Lester W. Milbrath, *Envisioning A Sustainable Society: Learning Our Way Out* (Albany: State Univ. of New York Press, 1989).

3   Ehrlich and Ehrlich, *Extinction*, 10. These authors note that "the rate at which populations and species are being eradicated has reached the point where a society like the United States would be better off doing without even the most 'necessary' projects if they cannot be carried out without causing further extinctions" (11). For similar views about preserving species, see David W. Ehrenfeld, *Conserving Life on Earth* (New York: Oxford Univ. Press, 1972), 4; and Charles Hartshorne, "The Rights of the Subhuman World," *Environmental Ethics* 1 (1979): 49–60.

4   David W. Ehrenfeld, *The Arrogance of Humanism* (New York: Oxford Univ. Press, 1978), 208.

5   Ehrlich and Ehrlich, *Extinction*, 11. Despite the Ehrlichs' warning, they admit they "would be sorely tempted to kiss the *Anopheles* goodbye and attempt to deal with . . . other problems if and when they arose" (12). For an effective argument against the demise of smallpox, see Bernard Dixon, "Smallpox—Imminent Distinction, and an Unresolved Dilemma," *New Scientist* 69 (26 Feb. 1976): 430–32. For more recent discussions, see Harold M. Schmeck, "Last Samples of Smallpox Pose a Quandary," *New York Times*, 3 Nov. 1987, C1, C3; and Barrie Penrose, "Smallpox Virus to be Destroyed," *Sunday Times* (London), 8 Oct. 1989, A4.

6   The Occupational Safety and Health Act of 1970, P.L. 91–596, Section 6(b)(5) and the Clean Air Act Amendments of 1977, P.L. 95–95, Section 108(a)(2).

7   Norman Myers, *The Sinking Ark* (Oxford: Pergamon, 1979), 51. For a similar view, see Margery L. Oldfield, *The Value of Conserving Genetic Resources* (Washington: NPS, 1984), 287.

8   CEQ, *Environmental Quality, 1980* (Washington, 1981), 32. According to the CEQ, "Genetic diversity is the amount of genetic variability among individuals in a single species, whether the species exists as a single interbreeding group or as a number of populations, strains, breeds, races, or subspecies." For the use of genetic diversity as a ranking criterion, see Donald H. Regan, "Duties of Preservation," in Bryan G. Norton, ed., *The Preservation of Species: The Value of Biological Diversity* (Princeton, N.J.: Princeton Univ. Press, 1986), 212; and Robert Allen, *How to Save the World: Strategy for World Conservation* (Totowa, N.J.: Barnes & Noble, 1980), 25.

9   Paul R. Ehrlich, Anne H. Ehrlich, and John P. Holdren, *Ecoscience: Population,*

Resources, Environment (San Francisco: W. H. Freeman, 1977), 345; CEQ, *Environmental Quality, 1980*, 34.

10 Department of State, *Proceedings of the U.S. Strategy Conference on Biological Diversity* (Washington, 1982), 24. See also CEQ, *Environmental Quality, 1980*, 50–52. Ireland's reliance on a genetically uniform strain of potatoes proved disastrous in 1845 as it faced an infamous blight fungus.

11 Myers, *Ark*, 34.

12 Ibid., 33. For further discussion of K- and r-selected species, see Joseph M. Moran, Michael D. Morgan, and James H. Wiersma, *Introduction to Environmental Science*, 2nd ed. (New York: W. H. Freeman, 1986), 103–5.

13 Myers, *Ark*, 50.

14 James E. Estes, Norman S. Smith, and John F. Palmisane, "Sea Otter Predation and Community Organization in the Western Aleutian Islands, Alaska," *Ecology* 59 (1978): 822–33; and David O. Duggins, "Kelp Beds and Sea Otters: An Experimental Approach," *Ecology* 61 (1980): 447–53.

15 Myers, *Ark*, 51. For informative introductions to the topic of bioaccumulation, see Ehrlich and Ehrlich, *Extinction*, 83–85; and Moran et al., *Environmental Science*, 50–52, 224–29.

16 For a discussion of trophic levels, see Richard Brewer, *Principles of Ecology* (Philadelphia: W. B. Sanders, 1979), 122–29.

17 Peter H. Raven, "Ethics and Attitudes," in J. B. Simmons et al., eds., *Conservation of Threatened Plants* (New York: Plenum, 1978).

18 Rice Odell, "Humans Attach Many Values to Wildlife Species," *Conservation Foundation Letter* (May 1978): 2. See also Paul Ehrlich, "Humankind's War against Homo Sapiens," *Defenders* 60 (Nov.-Dec., 1985), 5.

19 David W. Ehrenfeld, *Conserving Life*, 200–203; and John Terborgh, "Preservation of Natural Diversity: The Problem of Extinction Prone Species," *BioScience* 24 (Dec. 1974): 715–22.

20 Myers, *Ark*, 43–44. In a triage system there is obviously a third category—those that are likely to survive without attention. For endangered species this category makes no sense; if a species is likely to survive without attention, then it is not in danger of extinction.

21 Edwin Smith, "The Endangered Species Act and Biological Conservation," *Southern California Law Review* 57 (1984): 402.

22 Erik Eckholm, *Disappearing Species: The Social Challenge* (Washington: Worldwatch Institute, 1978), 25. Ironically, one of the first scientists to apply the triage analogy to the protection of endangered species is also one of its critics. Thomas Lovejoy discussed triage in "We Must Decide Which Species Will Go Forever," *Smithsonian* 7 (July 1976): 52–59. Ten years later, Lovejoy declared that "triage is both unworkable and misleading in its apparent common sense." See Lovejoy, "Species Leave the Ark One by One," in Norton, *Preservation of Species*, 23.

23 Mark Sagoff, "On the Preservation of Species," *Columbia Journal of Environmental Law* 7 (1980): 48. Others supporting a scheme that favors preservation of ecological communities include Bryan G. Norton, "Epilogue," in Norton, *Preservation of Species*, 279–83; and Ehrlich and Ehrlich, *Extinction*, 219–38.

24 Myers, *Ark*, 54–55.

25 Holmes Ralston, "Duties to Endangered Species," *BioScience* 35 (Dec. 1985): 720–

21; A. B. Shaw, "Adam and Eve, Paleontology, and the Non-objective Arts," *Journal of Paleontology* 43 (1969): 1085–98; and Bayard Webster, "Classification Is More Than a Matter of Fish or Fowl," *New York Times*, 14 Feb. 1982, sec. 4, 8.

26 Wilbur R. Jacobs, "Revising History with Ecology," in Roderick Nash, ed., *Environment and Americans* (New York: Holt, Rinehart, 1972), 85; and Continental Group, *Toward Responsible Growth* (Stamford, Conn.: Continental Group, 1982), 48. In its national survey of over 1,300 Americans, the Continental Group reported that 39 percent of the respondents disagreed with this statement: "Plants and animals *do not* exist primarily to be used by humans."

27 John Passmore, *Man's Responsibility for Nature: Ecological Problems and Western Traditions* (New York: C. Scribner's Sons, 1974), 5.

28 Talbot Page, *Conservation and Economic Efficiency: An Approach to Materials Policy* (Baltimore: Johns Hopkins Univ. Press, 1977), 9; and Julian L. Simon, *The Ultimate Resource* (Princeton, N.J.: Princeton Univ. Press, 1981), 149.

29 BSFW, *Draft Environmental Statement, Proposed Endangered Species Conservation Act of 1972*, DES #72–44 (Washington, 21 Mar. 1972).

30 J. R. Stoll and L. A. Johnson, "Concepts of Value, Nonmarket Valuation, and the Case of the Whooping Crane," *Transactions of the 49th North American Wildlife and Natural Resources Conference*, (Washington: Wildlife Management Institute, 1984), 382–93; K. J. Boyle and R. C. Bishop, "The Total Value of Wildlife: A Case Study Involving Endangered Species," paper presented at a joint meeting of the American Agricultural Economics Association and the Association of Environmental and Resource Economists, Aug. 1985, Ames, Iowa.

31 Ehrenfeld, *Arrogance of Humanism*, 201.

32 Colin W. Clarke, "Bioeconomics of the Ocean," *BioScience* 31 (Mar. 1981): 235. Clarke's purpose was to demonstrate that economic analysis can be used to justify extinction, not necessarily that such extinction is desirable. See also Clarke, "Profit Maximization and the Extinction of Animal Species," *Journal of Political Economy*, 81 (July-Aug., 1973): 950–61. Frank T. Bachmura suggests that economic analysis can be used to determine whether a species has a positive or negative present value. See "The Economics of Vanishing Species," *Natural Resource Journal* 11 (Oct. 1971): 674–92. For an economist's argument against the use of cost-benefit analysis, see Richard C. Bishop, "Endangered Species: An Economic Perspective," *Transactions of the 45th North American Wildlife and Natural Resources Conference*, (Washington: Wildlife Management Institute, 1980), 208–18.

33 Making judgments about relative values in the future also presupposes that future generations share this generation's discount rate as well as its preferences for the use of resources. Also, "a decision by the present generation not to conserve a species permanently denies future generations any benefits that the species may have provided in the future." Jane O. Yager, *An Application of Economic Efficiency Theory to Species Conservation* (Ph.D. Diss., Univ. of Maryland, 1986), 100.

34 Ehrenfeld, *Arrogance of Humanism*, 179, 189.

35 James G. March and Herbert A. Simon, *Organizations* (New York: Wiley, 1958), 140–41.

5 The Listing of Species, Organizational Choice, and Decision Rules

1   BSFW, *Rare and Endangered Fish and Wildlife of the United States* (Washington, 1966).

2   Steven Yaffee, *Prohibitive Policy* (Cambridge: MIT Press, 1982), 35.

3   Endangered Species Preservation Act of 1966, P.L. 89–669, Section 1(c).

4   The Endangered Species Preservation Act of 1966 formally created the National Wildlife Refuge System by including in it "all lands, waters, and interests therein administered by the Secretary [of the Interior] as wildlife refuges, areas for the protection and conservation of fish and wildlife that are threatened with extinction, wildlife ranges, game ranges, wildlife management areas, or waterfowl production areas. . . ." Section 4(c).

5   Michael Bean, *The Evolution of National Wildlife Law*, rev. ed. (New York: Praeger, 1983), 321.

6   The Land and Water Conservation Fund Act of 1965, P.L. 88–578, established a trust fund with revenues to be collected from user fees, the sale of excess federal property, and a tax on motorboat fuels. Subsequent amendments to the law added other revenue sources, including money from leases of federally owned offshore oil areas. Although the LWCF Act preceded the passage of the 1966 endangered species law, the former allowed the purchase of "any national area which may be authorized for the preservation of species . . . that are threatened with extinction." Section 6(a)(1).

7   *FR* 32 (11 Mar. 1967): 4001.

8   CREWS, *Rare and Endangered Fish and Wildlife of the United States* (Washington, 1968).

9   Endangered Species Conservation Act of 1969, P.L. 91–135, Section 3(a).

10   House Committee on Merchant Marine and Fisheries, hearings, *Predatory Mammals and Endangered Species*, 92nd Cong., 2nd sess., 1972, 117.

11   Ibid., 142.

12   House Committee on Appropriations, hearings, *Department of the Interior and Related Agencies Appropriations for 1974, Part 2*, 93rd Cong., 1st sess., 1973, 655.

13   House Committee on Merchant Marine and Fisheries, *Endangered and Threatened Species Conservation Act of 1973*, 93rd Cong., 1st sess., 1973, H. Rept. 93–412, 2.

14   House Committee on Appropriations, *Department of the Interior and Related Agencies Appropriations Bill, 1973*, 92nd Cong., 2nd sess., 1972, H. Rept. 92–1119, 21.

15   As Michael Bean points out, this is unlikely to be a meaningful provision: "the closer any insect pest approaches the brink of extinction, the risk it poses to man will almost necessarily be reduced in significance." Bean, *National Wildlife Law*, 332.

16   Comment of Lynn Greenwalt in House Committee on Appropriations, hearings, *Department of the Interior and Related Agencies Appropriations for 1975, Part 4*, 93rd Cong., 2nd sess., 1974, 483.

17   H. Rept. 93–412, *Endangered and Threatened Species Conservation Act of 1973*, 11.

18   Senate Committee on Commerce, hearings, *Endangered Species Conservation Act of 1972*, 92nd Cong., 2nd sess., 1972, 71.

19   The authorization for emergency protection of plants was added in 1979. Endangered Species Act of 1973, Appropriation Authorization, P.L. 96–159, Section

3(5). The same section extended the period of temporary protection to 240 days. The emergency provision has rarely been invoked, at least through early 1990.

20 Endangered Species Act of 1973, P.L. 93–205, Sections 3(4) and 3(15).

21 Senate Committee on Commerce, *Endangered Species Act of 1973*, 93rd Cong., 1st sess., 1973, S. Rept. 93–307, 7.

22 Endangered Species Act of 1973, P.L. 93–205, Section 10(b). This section allowed the granting of exemptions to those who would suffer "undue economic hardship" as a result of a decision to list a species. Section 10(e) allowed exceptions to the prohibition on taking to Alaskan natives as long as the taking was for subsistence purposes.

23 *FR* 41 (16 June 1976), 24525.

24 H. Rept. 93–412, *Endangered and Threatened Species Conservation Act of 1973*, 8. There is also some indication that foreign commerce in plants was prohibited not because congressmen believed it desirable, but rather in order to comply with the cites, which the Senate was about to ratify. See House Committee on Merchant Marine and Fisheries, *Endangered Species Act of 1973*, 93rd Cong., 1st sess., 1973, H. Rept. 93–740, 26.

25 Janet H. Weinberg, "Botanocrats and the Fading Flora," *Science News* 108 (9 Aug. 1975): 94.

26 Jeffery L. Pressman and Aaron Wildavsky, *Implementation*, 3rd ed. (Berkeley and Los Angeles: Univ. of California Press, 1984), 143.

27 House Committee on Merchant Marine and Fisheries, hearings, *Endangered Species Oversight*, 94th Cong., 1st sess., 1975, 15–16.

28 Smithsonian Institution, *Endangered and Threatened Plants of the United States*, reprinted as H. Doc. 94–51, 94th Cong., 1st sess., 1975. Vascular plants are those that have fluid-conveying vessels or ducts. The Smithsonian Institution's definitions of endangered and threatened paralleled those in the Endangered Species Act of 1973. Recently extinct or possibly extinct species represented "those species of plants no longer known to exist after repeated search of the type localities and other known or likely places." A species, though cultivated in gardens, was considered extinct if not found in the wild.

29 Ibid., 29.

30 *FR* 40 (1 July 1975): 27824.

31 House, *Endangered Species Oversight*, 282.

32 *FR* 41 (16 June 1976): 24524–72. Using criteria similar to those found in the Endangered Species Act, the Smithsonian Institution had listed 1,400 plants as endangered. Consequently, some plants that the Institution had classified as threatened or possibly extinct were included in the fws proposal, since some of these had been rediscovered. In addition to the Smithsonian list, the June proposal included approximately fifty plant species brought to the Service's attention after publication of the original list. For the remaining species on the Institution's list, the fws apparently concluded that the data did not meet the law's criteria for listing.

33 Yaffee, *Prohibitive Policy*, 135.

34 *ESTB* 1 (July 1976): 1. Schreiner provided even higher estimates in an article published in *Science News*. In it he estimated that the "endangered species universe has about 2 million species of plants and animals, give or take 100,000.

There are probably four or five times that many subspecies and God knows how many populations." See Weinberg, "Botanocrats," 92.

35 House, *Endangered Species Oversight*, 21.

36 Endangered Species Act of 1973, Section 7. Federal agencies must also consult with the NMFS when their projects potentially affect the relatively few species under its authority (see ch. 9).

37 Weinberg, "Botanocrats," 94.

38 Letter from Nathaniel P. Reed, assistant secretary for fish and wildlife and parks, to Lynn Seeber, general manager, TVA, 7 Mar. 1975. The petition was dated 20 Jan. 1975.

39 Letter from Lynn Seeber, general manager, TVA, to Nathaniel P. Reed, assistant secretary for fish and wildlife and parks, 12 Mar. 1975.

40 *FR* 40 (9 Oct. 1975): 47505–6.

41 For designation of critical habitat, see *FR* 41 (1 Apr. 1976): 13926–28. For petition to delist the habitat, see letter from Lynn Seeber, general manager, TVA, to Lynn Greenwalt, director, FWS, 28 Feb. 1977.

42 *Hill v.TVA*, 419 F.Supp. 753 (E.D. Tenn. 1976).

43 *TVA v. Hill*, 549 F.2d 1064 (6th Cir. 1977). The Court of Appeals' original injunction, issued in July 1976, halted construction, but was soon modified to allow construction that would not lead to the dam's closure, the action that supposedly would lead to the darter's demise.

44 The year after the snail darter was listed, as an example, the Senate Appropriations Committee directed that the dam "be completed as promptly as possible in the public interest." See Senate Committee on Appropriations, *Public Works for Water and Power Development and Energy Research Appropriation Bill, 1977*, 94th Cong., 2nd sess., 1976, S. Rept. 94–960, 96.

45 *TVA v. Hill*, 437 U.S. 153 (1978). For Griffin Bell and the snail darter, see Philip Hager, "Bell Asks Court to Rule Against Snail Darter," *Washington Post*, 19 Apr. 1978, A9.

46 Senate Committee on Environment and Public Works, hearings, *Amending the Endangered Species Act of 1973*, 95th Cong., 2nd sess., 1978, 329.

47 Ibid., 227.

48 Ibid., 45.

49 *Congressional Record*, 95th Cong., 2nd sess., 1978, 124, pt. 16: 21356.

50 Parenthetically, in passing the Endangered Species Act Amendments of 1978, Congress also indicated its appraisal of the protection of endangered species. The Endangered Species Act Authorization of 1976 (P.L. 94–325) provided funding for the program only until 30 September 1978. President Carter did not sign the 1978 law until 10 November, meaning that the program was without money and legislative authorization for nearly six weeks between 1 October and 10 November.

51 Endangered Species Act Amendments of 1978, P.L. 95–632, Section 11(1). Only in "rare circumstances" would the specification of critical habitat not be done concurrently with the proposed listing. See House Committee on Merchant Marine and Fisheries, *Endangered Species Act Amendments of 1978*, 95th Cong., 2nd sess., 1978, H. Rept. 95–1625, 17.

52 Endangered Species Act Amendments of 1978, P.L. 95–632, Section 2(2). Critical

habitats could also include "specific areas outside the geographical area occupied by the species at the time it is listed . . . upon a determination . . . that such areas are essential for the conservation of the species." Except in instances in which it was deemed necessary, critical habitats were not to include "the entire geographical area which can be occupied by the threatened or endangered species."

53   Ibid., Section 11(7). One exception to the benefit rule existed. No exclusions would be allowed if the "failure to designate such area as critical habitat [would] result in the extinction of the species." The House's version of the final bill would have allowed alteration of the critical habitat designation for invertebrate species only. See H. Rept. 95–1625, *Endangered Species Act Amendments of 1978*, 16.

54   There is also a third requirement, namely that "the benefits of such action clearly outweigh the benefits of alternative courses of action consistent with conserving the species or its critical habitat, and such action is in the public interest." See Endangered Species Act Amendments of 1978, P.L. 95–632, Section 7(h)(1)(A)(i), (ii), and (iii). In order to grant an exemption at least five votes would be necessary. The Endangered Species Act Amendments of 1982, P.L. 97–304, abolished the review board and transferred its functions and responsibilities to the secretary of the interior.

55   "Government Panel Confirms TVA Dam Must Give Way to Survival of Fish," *Wall Street Journal*, 24 Jan. 1979, 2. Few applications for exemptions have ever been filed. The Pittston Company initiated one request in early 1979 in regard to its plans to build an oil refinery and marine terminal at Eastport, Maine. The request was later withdrawn and the consultation process was reinitiated. See *ESTB* 4 (Mar. 1979): 1, 3. The Consolidated Grain and Barge Company of St. Louis, Missouri, filed an application for an exemption involving the orange-footed pearly mussel in late 1985, but the company subsequently decided not to pursue its request. See *FR* 50 (6 Nov. 1985): 46157.

56   Energy and Water Development Appropriation Act of 1980, P.L. 96–69, Title IV, (1979).

6   Extending the List of Species in Jeopardy

1   *FR* 44 (6 Mar. 1979): 12382–83.
2   *FR* 44 (15 Aug. 1979): 47862.
3   *ESTB* 5 (Jan. 1980): 1, 3–4.
4   Philip Shabecoff, "Interior Department Assailed on Missing a Deadline for Species Protection," *New York Times*, 7 Nov. 1979, A23.
5   Endangered Species Act Amendments of 1978, P.L. 95–632, Section 11(5).
6   GAO, *Endangered Species—A Controversial Issue Needing Resolution* (Washington, 1979). Although formally released on 2 July 1979, Congress had been appraised of the report's findings several months earlier. It should be noted that the Interior Department disagreed with nearly all the GAO findings and claimed that GAO auditors did not have the biological competence to sustain its conclusions. Before the final report was issued, Interior provided the GAO with a rebuttal of more than one hundred pages. The GAO response to the rebuttal reveals rather contentious relations between it and Interior: "Some of Interior's comments were useful for making corrections. . . . However, other comments either were contradictory with

previous information received from Interior or other sources, irrelevant to the issues at hand, or inaccurate." Ibid., 121.

7   Statement of Robert Herbst in Senate Committee on Appropriations, hearings, *Department of the Interior and Related Agencies Appropriations, Fiscal Year 1980, Part 3*, 96th Cong., 1st sess., 1979, 2260; comments of Earl Baysinger, assistant chief, Office of Endangered Species and International Activities, in House Committee on Merchant Marine and Fisheries, hearings, *Endangered Species Oversight*, 94th Cong., 1st sess., 1975, 78.

8   Ibid.

9   Endangered Species Act of 1973, P.L. 93–205, Section 4(a).

10 GAO, *Endangered Species—A Controversial Issue*, 11. The dam affected the turgid-blossom pearly mussel, which was listed as endangered on 14 June 1976.

11 *FR* 43 (6 July 1978): 29152.

12 "Summary: Monsanto and the Illinois Mud Turtle," and letter to Clarence Thomas from W. D. Carpenter, director, Environmental Operations, Monsanto Agricultural Products Co., 16 Jan. 1980. Both the summary and Carpenter's letter are attachments to a letter from Senator John C. Culver (D.–Iowa) to Lynn Greenwalt, director, FWS, 22 Jan. 1980. A representative from Monsanto characterized FWS scientific data as deplorable and inadequate, "full of second-hand information [and] speculation. . . ." See House Committee on Merchant Marine and Fisheries, hearings, *Endangered Species*, 96th Cong., 1st sess., 1979, 267–68.

13 Letter from Roger Jepson to Cecil Andrus, secretary of the interior, 29 Jan. 1980. Jepson appended a handwritten note: "This letter is very low key—as compared to how I really feel."

14 The FWS position on the species that the New Melones dam in California would affect is somewhat difficult to ascertain. In its response to a draft copy of the GAO report, Interior stated at the end of 1978 that "insufficient field work had been done to justify proposing the Melones cave harvestman as an Endangered Species." Attachment to letter from Meierotto to Eschwege, GAO, 17 Apr. 1979, 9. Soon after the release of the GAO report in July 1979, however, the FWS concluded that the cave harvestman should be listed as threatened. No action was ever taken to do so. See House, *Endangered Species*, 218.

15 GAO, *Endangered Species—A Controversial Issue*, 15.

16 Attachment to letter from Meierotto to Eschwege, GAO, 17 Apr. 1979, 16. For the delisting, see *FR* 48 (22 Nov. 1983): 52740–43.

17 GAO, *Endangered Species—A Controversial Issue*, 16.

18 Ibid., 29.

19 Senate Committee on Appropriations, hearings, *Proposed Critical Habitat Area for Grizzly Bears*, 94th Cong., 2nd sess., 1976, 10.

20 Timothy Egan, "Ruling on Owl Stirs New Hope for Trees," *New York Times*, 11 Nov. 1988, A16; and GAO, *Endangered Species: Spotted Owl Petition Evaluation Beset by Problems* (Washington, 1989).

21 Resource Planning Associates, *Improving the Cost-Effectiveness of the Endangered Species Program* (Washington: Resource Planning Assoc., 1980), 34.

22 Lynn Greenwalt, FWS director for much of the 1970s, essentially agreed with this view when he noted: "There is also an element of misunderstanding about

what designation as threatened really means. The determination as to what it takes to be threatened is, in a way, a subjective value judgment. There is no clear guideline in the act that indicates just where the threatened category begins and another one leaves off." See Senate Committee on Interior and Insular Affairs, hearings, *Interior Nomination*, 93rd Cong., 2nd sess., 1974, 33.

23 *FR* 43 (23 Aug. 1978): 37662.

24 Ibid., 45 (20 Aug. 1980): 55655.

25 GAO, *Endangered Species—A Controversial Issue*, 24.

26 Ibid., 25.

27 Letter from John L. Paradiso, acting chief, OES, to Carl Koford, research zoologist, Univ. of California, Berkeley, 28 Oct. 1976.

28 GAO, *Endangered Species—A Controversial Issue*, 27.

29 Gene Ruhr, "Setting Priorities," unpublished manuscript, 1971.

30 Rollin D. Sparrowe and Howard M. Wight, "Setting Priorities for the Endangered Species Program," in *Transactions of the 40th North American Wildlife and Natural Resources Conference* (Pittsburgh: Wildlife Management Institute, 1975), 142–54.

31 FWS, "Appendix 1: Priority System," in *Endangered Species Program Management Document, 1980* (Washington, 1980), ii.

32 House, *Endangered Species*, 234.

33 Senate Committee on Environment and Public Works, hearings, *Endangered Species Act Oversight*, 97th Cong., 1st sess., 1981, 286.

34 Executive Order 12291, *FR* 46 (19 Feb. 1981): 13193.

35 Letter from G. Ray Arnett, assistant secretary for fish and wildlife and parks, to Rep. Edwin B. Forsythe (R.–New Jersey), 30 Apr. 1982.

36 House Committee on Merchant Marine and Fisheries, hearings, *Endangered Species Act*, 97th Cong., 2nd sess., 1982, 337.

37 Ibid., 160.

38 Enclosure with letter from Roy J. Kirk, senior group director, Community and Economic Development Division, GAO, to G. Ray Arnett, assistant secretary of the interior for fish and wildlife and parks, 1 Apr. 1982.

39 House Committee on Merchant Marine and Fisheries, *Endangered Species Act Amendments of 1982*, 97th Cong., 2nd sess., 1982, H. Rept. 97–835, 20.

40 Endangered Species Act of 1973, Appropriation Authorization, P.L. 96–159, Section 4(h)(3), 1979.

41 House, *Endangered Species Act*, 553, and FWS, Draft Report, "Appendix I: Listing and Recovery Priority Systems," (Washington, 1981), 34.

42 Attachment to letter from Meierotto to Eschwege, GAO, 17 Apr. 1979, 26. During one interview with an OES scientist involved with the listing process, the scientist commented that the FWS is more "bird, fish, and mammal oriented." He also said that "old-line Fish and Wildlife types, regional directors particularly, don't want to get involved with bugs." Interview at headquarters, OES, 21 Aug. 1986.

43 Cited in Roger Di Silvestro, "Our Looming Failure on Endangered Species," *Defenders* 59 (July-Aug. 1984): 27.

44 House, *Endangered Species Act*, 553.

45 H. Rept. 97–835, *Endangered Species Act Amendments of 1982*, 21. In this report the

conference committee instructed the FWS to "utilize a scientifically based priority system. . . . Distinctions based on whether the species is a higher or lower life form are not to be considered."

46   *FR* 48 (21 Sept. 1983): 43103.
47   *FR* 45 (15 June 1980): 82480.
48   For vertebrates, see *FR* 47 (30 Dec. 1982), 58454–60; and corrections in *FR* 48 (17 May 1983): 22173–74. For plants, see *FR* 48 (28 Nov. 1983): 53640–70.
49   Statement of G. Ray Arnett, assistant secretary of the interior for fish and wildlife and parks, in Senate Committee on Appropriations, hearings, *Department of the Interior and Related Agencies Appropriations, Fiscal Year 1983, Part 3*, 97th Cong., 2nd sess., 1982, 342.
50   See House Committee on Appropriations, hearings, *Department of the Interior and Related Agencies Appropriations for 1982, Part 1*, 97th Cong., 1st sess., 1981, 976; and ibid., *Department of the Interior and Related Agencies Appropriations for 1983, Part 1*, 97th Cong., 2nd sess., 1982, 504.
51   FWS, OES, *Guide to Preparing and Processing Actions to List, Delist, or Reclassify Endangered or Threatened Species*, 2nd ed., (Washington, 1985).
52   Interview at headquarters, OES, 21 Aug. 1986.

7   Preserving Species and Protecting Habits

 1   Norman Myers, *The Sinking Ark* (Oxford: Pergamon, 1979), 225; Jared M. Diamond, "The Island Dilemma: Lessons of Modern Biogeographic Studies for the Design of Natural Reserves," *Biological Conservation* 7 (1975): 129–46; John Terborgh, "Preservation of Natural Diversity: The Problem of Extinction Prone Species," *BioScience* 24 (Dec. 1974): 715–22.
 2   William D. Newmark, "A Land-Bridge Island Perspective on Mammalian Extinctions in Western North American Parks," *Nature* 325 (29 Jan. 1987): 430–32.
 3   Myers, *Sinking Ark*, 224.
 4   107 S.Ct. 2378 (1987). See Al Kamen, "Land-Use Rights Are Bolstered," *Washington Post*, 10 June 1987, A1.
 5   In the 1970s a survey was administered to ten OES professional employees. Each employee was asked to rank different species on the basis of their economic, aesthetic, social, and ecological value. Mammals were ranked highest, followed by fishes, birds, and reptiles/amphibians. See Gardner Brown, Jr., "Survey of Methodologies for Ranking Species," revised report prepared for the Department of the Interior, Office of Policy Analysis, May 1982, 12.
 6   Terborgh, "Preservation of Natural Diversity," 719.
 7   PLLRC, *One Third of the Nation's Land* (Washington, 1970), 160.
 8   "Take" means to "harass, harm, pursue, hunt, shoot, wound, kill, trap, capture, collect, or attempt to engage in any such conduct." Endangered Species Act of 1973, P.L. 93–205, Section 3(14).
 9   Endangered Species Preservation Act of 1966, P.L. 89–669, Section 1(b).
10   Ibid., Section 2(c).
11   GAO, *Endangered Species—A Controversial Issue Needing Resolution* (Washington, 1979), 111–12; and The Nature Conservancy, *Preserving Our Natural Heritage* (Washington: The Nature Conservancy, 1976), 1:67. Interior also acquired 462

acres at a cost of $936,000 to establish a research facility at Patuxent, Maryland.

12 Senate Committee on Appropriations, hearings, *Department of the Interior and Related Agencies Appropriations, Part 3*, 91st Cong., 2nd sess., 1970, 2857.

13 This determination of threat was made in 1978, so the prior acquisition of land may have diminished the threat to the species involved (i.e., the key deer, the Columbian white-tailed deer, and the Delmarva Peninsula fox squirrel).

14 House Committee on Merchant Marine and Fisheries, *Endangered and Threatened Species Conservation Act of 1973*, 93rd Cong., 1st sess., 1973, H. Rept. 93–412, 5; Endangered Species Act of 1973, P.L. 93–205, Section 2(b).

15 Ibid., Section 7.

16 Ibid., Section 3(1).

17 Donald S. Van Meter and Carl E. Van Horn, "The Policy Implementation Process: A Conceptual Framework," *Administration and Society* 6 (Feb. 1975): 445–87.

18 As one spokesman for the Nixon administration testified, the section prohibiting jeopardy to listed species would be "the first piece of substantive law which agencies would have to adhere to in carrying out their programs and duties. . . ." See House Committee on Merchant Marine and Fisheries, hearings, *Endangered Species*, 93rd Cong., 1st sess., 1973, 188.

19 National Environmental Policy Act of 1969, P.L. 91–190, Section 102(2)(c).

20 Serge Taylor, *Making Bureaucracies Think: The Environmental Impact Statement Strategy of Administrative Reform* (Stanford, Calif.: Stanford Univ. Press, 1984), 7.

21 Walter Rosenbaum, *The Politics of Environmental Concern*, 2nd ed. (New York: Praeger, 1977), 119.

22 See H. Rept. 93–412, *Endangered and Threatened Species Conservation Act of 1973*, 14; and House Committee on Merchant Marine and Fisheries, *Endangered Species Act of 1973*, 93rd Cong., 1st sess., 1973, H. Rept. 93–740.

23 Martha T. Eider-Orley, "The Affirmative Duty of Federal Departments and Agencies to Restore Endangered and Threatened Species," *Hofstra Law Review* 6 (1978): 1075.

24 House Committee on Merchant Marine and Fisheries, hearings, *Endangered Species*, 96th Cong., 1st sess., 1979, as cited in Cathryn Campbell, "Federal Protection of Endangered Species: A Policy of Overkill?" *Journal of Environmental Law* 3 (1983): 254. These are the two sentences:

> The Secretary shall review other programs administered by him and utilize such programs in furtherance of the purposes of this Act. All other departments and agencies shall, in consultation with and with the assistance of the Secretary [of Commerce or of the Interior] utilize their authorities in furtherance of the purposes of this Act by carrying out programs for the conservation of endangered species and threatened species . . . and by taking such action necessary to insure that actions authorized, funded, or carried out by them do not jeopardize the continued existence of such endangered species and threatened species or result in the destruction or modification of habitat of such species which is determined by the Secretary, after consultation as appropriate with the affected States, to be critical.

25 Francis E. Rourke, *Bureaucracy, Politics, and Public Policy*, 3rd ed. (Boston: Little, Brown, 1984), 101.

26 Graham T. Allison, *Essence of Decision* (Boston: Little, Brown, 1971), 84.

27 Eugene Bardach and Robert A. Kagan, *Going By the Book: The Problem of Regulatory Unreasonableness* (Philadelphia: Temple Univ. Press, 1982), 43.

28 Matthew Holden Jr., *Pollution Control as a Bargaining Process* (Ithaca, N.Y.: Water Resources Center, Cornell Univ., 1966), 11.

29 *FR* 40 (22 Apr. 1975): 17764.

30 Statement of Keith M. Schreiner in *ESTB* 1 (Aug. 1976): 1.

31 *FR* 40 (22 Apr. 1975): 17765.

32 Statement of Lynn Greenwalt in House Committee on Merchant Marine and Fisheries, hearings, *Endangered Species Oversight*, 94th Cong., 1st sess., 1975, 259.

33 *FR* 40 (16 May 1975): 21499–501.

34 FWS/NMFS, "Guidelines to Assist Federal Agencies in Complying with Section 7 of the Endangered Species Act of 1973," (22 Apr. 1976). The FWS and the NMFS had convened an ad hoc interagency committee of representatives from eleven federal agencies to advise the two services in the development of the guidelines.

35 Ibid.

36 The guidelines defined "jeopardy" to include "any action which reasonably would be expected to result in the reduction of the reproductive ability, numbers, or distribution of a listed species to such an extent that the loss would pose a threat to the continued survival or recovery of these species in the wild." Destruction or adverse modification would "include any act which would have a deleterious effect upon any of the constituent elements of critical habitat which are necessary to the survival or recovery of [listed] species, and such effect is likely to result in a decline in the numbers of the species."

37 Holden, *Pollution Control*, 37.

38 FWS/NMFS, "Guidelines."

39 Campbell, "Federal Protection," 255.

40 Ibid.

41 FWS/BLM/BOR, "Memorandum of Understanding on Interagency Coordination in Non-Emergency Critical Habitat Determinations Pursuant to Section 7 of the Endangered Species Act of 1973," (Mar. 1976).

42 Edwin C. Hargrove, *The Missing Link: The Study of the Implementation of Social Policy* (Washington: Urban Institute, 1975), 113.

43 Statement of Keith M. Schreiner in Senate Committee on Environment and Public Works, hearings, *Endangered Species Act Oversight*, 95th Cong., 1st sess., 1977, 381.

44 *ESTB* 2 (Dec. 1976–Jan. 1977): 3; John J. Craighead and John A. Mitchell, "Grizzly Bear," in J. A. Chapman and G. A. Feldhamer, eds., *Wild Mammals of North America* (Baltimore: John Hopkins Univ. Press, 1982), 533.

45 FWS Region 6, "Draft Environmental Assessment: Proposed Determination of Critical Habitat for the Grizzly Bear of the 48 Conterminous States," (June 1976), 1; *FR* 41 (5 Nov. 1976): 48757–59.

46 *ESTB* 1 (Aug. 1976): 3.

47 FWS, "Environmental Assessment . . . for the Grizzly," 13.

48 Ibid., 12.

49 Senate Committee on Appropriations, hearing, *Proposed Critical Habitat Area for Grizzly Bears*, 94th Cong., 2nd sess., 1976.

50   Cited in George Laycock, "Uproar over Grizzly Habitat," *Audubon* 79 (May 1977): 126.

51   Memorandum of David B. Marshall, senior staff biologist, FWS Region 1, Portland, Oregon, in regard to "Hearings on Grizzly Bear Critical Habitat Proposal at St. Anthony, Idaho, on December 14, 1976," 21 Dec. 1976.

52   Memorandum from acting associate director, FWS, to regional director, FWS Region 2, Albuquerque, "Houston Toad Critical Habitat," 7 Dec. 1976.

53   Memorandum from acting regional director, FWS Region 2, Albuquerque, to director, FWS, "Reply to Comments on Proposed Critical Habitat for the Houston Toad," 20 Dec. 1976.

54   *FR* 42 (26 May 1977): 27009.

55   Memorandum from James E. Johnson, endangered species specialist, FWS Region 2, Albuquerque, to regional director, FWS Region 2, Albuquerque, 25 Nov. 1977. These views received an important endorsement from the director of the wildlife division of the Texas Parks and Wildlife Department, Ted L. Clark. He favored all but one site in Harris County and recommended that another area be enlarged. See *FR* 43 (31 Jan. 1978): 4022.

56   Memorandum from R. F. Stephens, acting regional director, FWS Region 2, Albuquerque, to director, FWS 25 Nov. 1977.

57   *ESTB* 3 (Feb. 1978): 1. For final designation of the critical habitat, see *FR* 43 (31 Jan. 1978): 4022.

58   Memorandum from Keith M. Schreiner, associate director, FWS, to Service Directorate and chiefs of divisions and offices, 2 Mar. 1978.

59   Memorandum from assistant regional director, FWS Region 2, Albuquerque, to director, FWS, 30 Mar. 1978.

60   Steven Yaffee, *Prohibitive Policy* (Cambridge: MIT Press, 1982), 96.

61   Statement of John W. Grandy, Defenders of Wildlife, in Senate Committee on Commerce, hearings, *To Amend the Endangered Species Act of 1973*, 94th Cong., 2nd sess., 1976, 107. When the final rule designating 800 square miles as critical habitat for the condor was published, the *Los Angeles Times* reported that the habitat was "not to be disturbed by man." "'Critical Habitat' for Condors Declared," *Los Angeles Times*, 23 Sept. 1976, sec. 1, 2. For the butterflies, see "Government Seeks Refuge for Rare Butterflies," *Los Angeles Times*, 4 Feb. 1977, sec. 1, 27.

62   *ESTB* 2 (June 1977): 1–2. President Carter's memorandum is reprinted in House Committee on Merchant Marine and Fisheries, hearings, *Endangered Species— Part 1*, 95th Cong., 2nd sess., 1978, 95–96.

63   House Committee on Merchant Marine and Fisheries, *Endangered Species Act Amendments of 1978*, 95th Cong., 2nd sess., 1978, H. Rept. 95–1625, 13.

8   The Costs and Consequences of Experience

1   *FR* 43 (4 Jan. 1978): 876.

2   *FR* 42 (26 Jan. 1977): 4868. The FWS and the NMFS jointly issued the proposed regulations.

3   *National Wildlife Federation* v. *Coleman*, 529 F.2d 359 (1976); and *TVA* v. *Hill*, 549 F.2d 1064 (1977).

4 *FR* 43 (4 Jan. 1978): 870–76.

5 Ibid., 872.

6 Ibid.

7 Ibid., 875. The definition of critical habitat provided in 1978 was slightly different than the one offered in 1976. Critical habitat was defined in the 1978 regulations as "any air, land, or water area (exclusive of those existing manmade structures or settlements which are not necessary to the survival and recovery of the listed species) and constituent elements thereof, the loss of which would appreciably decrease the likelihood of the survival and recovery of a listed species or a distinct segment of its population." The definitions of jeopardy and destruction or adverse modification were changed only slightly from the earlier 1976 guidelines.

8 House Committee on Merchant Marine and Fisheries, *Endangered Species Act Authorization*, 94th Cong., 2nd sess., 1976, H. Rept. 94–887, 3.

9 Statement of John W. Grandy, in Senate Committee on Commerce, hearings, *To Amend the Endangered Species Act of 1973*, 94th Cong., 2nd sess., 1976, 106.

10 GAO, *Endangered Species—A Controversial Issue Needing Resolution* (Washington, 1979), 40. The three species included the California condor, the blunt-nosed leopard lizard, and the San Joaquin kit fox.

11 Ibid., 42.

12 Winston Harrington, *Endangered Species Protection and Water Resource Development* (Los Alamos, N.M.: Los Alamos Scientific Laboratory, 1980), 30. The legal action involved *Nebraska et al.* v. *Rural Electrification Administration, Environment Reporter: Cases* 12 (1978): 1156. This case involved the Grayrocks Dam and Reservoir, a water storage facility for three steam-electric generating units to be constructed in southeastern Wyoming. One concern of the plaintiffs was that the dam would reduce water flows in the Platte River in Nebraska, thus affecting the whooping crane's habitat. In addition to the state of Nebraska, the plaintiffs included the National Audubon Society and the National Wildlife Federation. The court's decision was issued in October 1978, and the Endangered Species Act Amendments of 1978 required a special committee to consider an exemption to the project. The competing sides reached a compromise thus eliminating the need for a decision from the committee.

13 GAO, *Endangered Species*, 49.

14 House Committee on Merchant Marine and Fisheries, *Endangered Species Act Amendments of 1978*, 95th Cong., 2nd sess., 1978, H. Rept. 95–1625, 11.

15 Senate Committee on Environment and Public Works, hearings, *Amending the Endangered Species Act of 1973*, 95th Cong., 2nd sess., 1978, 31.

16 Comment of Rep. David R. Bowen (D.–Mississippi) *Congressional Record*, 95th Cong., 2nd sess., 1978, 124, pt. 28: 38131.

17 H. Rept. 95–1625, *Endangered Species Act Amendments of 1978*, 13.

18 Senate Committee on Environment and Public Works, *Endangered Species Act Amendments of 1978*, 95th Cong., 2nd sess., 1978, S. Rept. 95–874, 2. The House Committee on Merchant Marine and Fisheries expressed similar concerns: "It is clear . . . that there will continue to be some Federally authorized activities which cannot be modified in a manner which will avoid a conflict with a listed species." H. Rept. 95–1625, *Endangered Species Act Amendments of 1978*, 13.

19 As the House Committee on Merchant Marine and Fisheries observed, "these provisions will ensure that the Department of the Interior is not . . . designating critical habitat without consulting the views of the people of the affected area." H. Rept. 95–1625, *Endangered Species Act Amendments of 1978*, 16. This is a curious statement because designations of critical habitats are supposed to be based on the best scientific data available, not on public opinion.

20 Endangered Species Act Amendments of 1978, P.L. 95–632, Section 7(a).

21 Endangered Species Act of 1973, Authorization Appropriation, P.L. 96–159, 1979. The 1979 law, which was intended primarily to authorize additional expenditures for the implementation of the Endangered Species Act, also required agencies to confer on any of their actions likely to jeopardize the continued existence of any species *proposed* for listing or critical habitat *proposed* for designation. For such proposals, action agencies would not face any limitation on their commitment of resources during the conference process as was the case with consultations on listed species and critical habitats already designated.

22 *FR* 48 (29 June 1983): 29990–30004 (proposed rules); and *FR* 51 (3 June 1986): 19926–63 (final rules). The FWS and the NMFS jointly developed the proposed and final rules.

23 *FR* 43 (4 Jan. 1978): 873.

24 *FR* 51 (3 June 1986): 19958. Emphasis added.

25 Graham T. Allison, *The Essence of Decision: Explaining the Cuban Missile Crisis* (Boston: Little, Brown, 1971), 83, 89.

26 FWS/NMFS, "Critical Habitat Priorities," 17 Feb. 1978.

27 Ibid.

28 These include the Maryland and fountain darters and the hawksbill and leatherback sea turtles. The actions for the turtles were joint designations involving both the FWS and the NMFS. The American peregrine falcon, which was in category 1, also had a critical habitat designated, but this was done in August 1977, six months before the priority list was issued. This suggests that the falcon was erroneously included on the priority list.

29 *FR* 44 (6 Mar. 1979): 12382.

30 H. Rept. 95–1625, *Endangered Species Act Amendments of 1978*, 17.

31 *FR* 51 (3 June 1986): 19927.

32 Interview at headquarters, OES, 21 Aug. 1986. See also John G. Sidle, "Critical Habitat Designation: Is It Prudent?" *Environmental Management* 11 (1987): 429–37.

33 House Committee on Merchant Marine and Fisheries, hearings, *Endangered Species Act*, 97th Cong., 2nd sess., 1982, 499.

34 *FR* 48 (11 Oct. 1983): 46055.

35 *ESTB* 9 (June 1984): 10.

36 *FR* 45 (25 Sept. 1980): 63812.

37 Lorraine Bennett, "Coachella Valley Gets a 'Snail Darter,'" *Los Angeles Times*, 3 Aug. 1980, sec. 2, 1, 3. The lizard continued to cause controversy several years after designation of its critical habitat. See Wesley G. Hughes, "Lizard May Halt Desert Growth," *Los Angeles Times*, 27 Dec. 1983, 1; and Ken Wells, "How a Lizard Gets Palm Springs to Do Its Bidding," *Wall Street Journal*, 9 July 1984, 1, 21.

38 *FR* 40 (16 May 1975): 21501.

39   GAO, *Obligations and Outlays from the Land and Water Conservation Fund* (Washington, 1986), 9. According to the Land and Water Conservation Act, not less than 40 percent of appropriated funds are for federal agencies' acquisition of land. The remainder is available, on a fifty-fifty matching basis, for the states. For a discussion of the LWCF, see Michael Mantell, Phyllis Myers, and Robert B. Reed, "The Land and Water Conservation Fund: Past Experiences, Future Directions," in William J. Chandler, ed., *Audubon Wildlife Report 1988/89* (San Diego: Academic Press, 1988), 257–81.

40   *ESTB* 5 (June 1980): 6.

41   GAO, *Endangered Species—A Controversial Issue*, 79.

42   Branch of Biological Support, OES, "Listed Species of U.S. and Territories Prioritized by P.M.D. Priority System and Recovery Status," Apr. 1978.

43   For an interesting discussion of how scientists tried to regenerate the species by crossbreeding the last males, see Kenny Gruson, "Effort to Save a Species: Brief Cheer, Then Gloom," *New York Times*, 7 Oct. 1985, A26; and John Noble Wilford, "Last Dusky Sparrow Struggles On," *New York Times*, 29 Apr. 1986, C1.

44   *ESTB* 5 (June 1980): 6.

45   Presidential Commission on Outdoor Recreation Resources Review, *Americans Outdoors: The Legacy, the Challenge* (Washington: Island Press, 1987).

46   Cass Peterson, "Outdoors Report Won't See the Light of Day," *Washington Post*, 6 Feb. 1987, A21. The lawsuit was based on an alleged violation of the Federal Advisory Committees Act. Republican reaction to the trust fund concept in the House was quite hostile. See George Hager, "Markup of Land-Purchase Bill Prompts Walkout by GOP," *Congressional Quarterly Weekly Report*, 6 May 1989, 1037–38.

47   Endangered Species Act of 1973, as amended, Section 7(a)(2).

48   Marc Reisner, *Cadillac Desert: The American West and Its Disappearing Water* (New York: Viking, 1986), 125.

49   Ibid.

50   R. J. Behnke and D. E. Benson conclude: "The most obvious and clearly identifiable factor contributing to the decline of [Colorado] squawfish is the large dams and reservoirs that converted hundreds of miles of large-river habitat into great impoundments. The preservation of native fishes was not considered in the planning and operation of these projects." See Behnke and Benson, *Endangered and Threatened Fishes of the Upper Colorado River Basin* (Fort Collins, Colo.: Cooperative Extension Service, Colorado State Univ., 1983), 13.

51   Testimony of Robert P. Davison, National Wildlife Federation, in Senate Committee on Environment and Public Works, hearings, *Endangered Species Act Authorizations*, 99th Cong., 1st sess., 1985, 163, 165.

52   See comments of James V. Smith, vice president, *The Waterways Journal*, in House Committee on Merchant Marine and Fisheries, hearings, *Endangered Species— Part 1*, 95th Cong., 2nd sess., 1978, 47; and Rep. James P. Johnson (R.–Colorado) in *Congressional Record*, 95th Cong., 2nd sess., 1978, 124, pt. 28: 38131.

53   BLM, *Moon Lake Project Draft Environmental Impact Statement*, 8 Jan. 1981, 14. Similarly, in another jeopardy opinion issued to the Dolores Project in Colorado in 1980, the FWS asserted that it was "important not to reduce present flows until we obtain sufficient biological data that insures that any reductions would not be

harmful." Cited in testimony of Robert P. Davison in Senate, *Endangered Species Act Authorizations*, 155.

54 Letter from Tony Willardson, associate director, Western States Water Council, to the author, 12 Aug. 1986.

55 Memorandum, "Biological Opinion for Windy Gap Project, Colorado," from director, FWS Region 6, Denver, to regional director, Lower Missouri Region, BLM, 13 Mar. 1981. In this opinion the FWS concluded that, without its proposed conservation measures, the project would be expected "to appreciably reduce the likelihood of the *recovery* of the endangered fish species." Emphasis in original. See also, GAO, *Endangered Species: Limited Effect of Consultation Requirements on Western Water Projects* (Washington, 1987), 28.

56 Although the Windy Gap decisions applied only to the Upper Colorado River Basin, there was a similar decrease in the number of jeopardy opinions issued for all western water projects after 1980. In the four years between January 1977 and December 1980, jeopardy opinions were issued in twenty of sixty-three formal consultations involving water projects in the West. Over the next four years, through the end of 1984, only five of sixty-two consultations (or 8 percent) resulted in jeopardy opinions. See Senate, *Endangered Species Act Authorizations*, 165.

57 GAO, *Endangered Species: Limited Effect of Consultation Requirements*, 29. One spokesman for the Western States Water Council interpreted the FWS choice in these terms: "Faced with growing discontent over delays to legitimate and necessary water projects (as compared to speculative development), the Service decided to compromise rather than adopt an obstructionist stance which might solidify support for amendments to the Act that would seriously weaken endangered species protections." See Willardson letter, 12 Aug. 1986.

58 Senate, *Endangered Species Act Authorizations*, 156; "Prepared Joint Statement of the Environmental Defense Fund, Sierra Club, Trout Unlimited, Friends of the Earth, and Colorado Audubon Council," in House Committee on Merchant Marine and Fisheries, hearings, *Endangered Species Act*, 99th Cong., 1st sess., 1985, 210.

59 GAO, *Endangered Species: Limited Effect of Consultation Requirements*, 27. The summary of the plan that follows is from the GAO report.

60 Defenders of Wildlife, *Saving Endangered Species: Implementation of the Endangered Species Act* (Washington: Defenders of Wildlife, 1987), 15.

61 Statement of Ronald Lambertson, associate director, Federal Assistance, FWS, in Senate Committee on Environment and Public Works, hearings, *Endangered Species Act Oversight*, 97th Cong., 1st sess., 1981, 22.

62 House, *Endangered Species Act*, 1982, 299–302. The calculations exclude situations in which jeopardy opinions were issued, but in which discussions were continuing at the time of the analysis.

63 OES, "Reauthorization of the Endangered Species Act, Section 7 Consultation Summaries FY 1982 to 1984," n.d.

64 For the least tern, see letter from Edward Collins, acting area manager, FWS Sacramento Field Office, to James C. Wolfe, chief, Construction-Operations Division, San Francisco District, Corps of Engineers, 21 July 1982 and letter from Jack Farless, chief, Construction-Operations Division, San Francisco Dis-

304 Notes

trict, Corps of Engineers, to the author, 25 Nov. 1986. For the seaside sparrow, see letter from James W. Pulliam, Jr., director, FWS Region 4, Atlanta, to A. J. Salem, chief, Planning Division, Corps of Engineers, Jacksonville District, 12 Oct. 1983; and letter from A. J. Salem, chief, Planning Division, Corps of Engineers, Jacksonville District, to the author, 16 Sept. 1986.

65 James Serfis, Richard Tinney, and Roger E. McManus, *The Environmental Protection Agency's Implementation of the Endangered Species Act with Respect to Pesticide Registration* (Washington: Center for Environmental Education, 1986), 21. Much of the discussion that follows is from this report.

66 Ibid., Appendix, 41.

67 Ibid., p. 37.

68 House Committee on Merchant Marine and Fisheries, *Endangered Species Act Amendments of 1988*, 100th Cong., 2nd sess., 1988, H. Rept. 100–928, 8–9.

69 Statement of John A. Moore, assistant administrator for pesticides and toxic substances, EPA, in Mary Connelly and Laura Mansnerus, "The EPA Takes a Look at Itself," *New York Times*, 31 Aug. 1986, sec. 4, 16.

70 G. Bruce Doern, *The Peripheral Nature of Scientific and Technological Controversy in Federal Policy Formation* (Ottawa: Science Council of Canada, 1981), 49.

71 House, *Endangered Species—Part 1*, 113.

72 House, *Endangered Species Act*, 1982, 409.

73 Doern, *The Peripheral Nature*, 24.

74 In testimony presented to the House Subcommittee on Wildlife Conservation and the Environment in early 1982, a spokesman from the National Wildlife Federation discussed the EPA's handling of registration for chlorpyfiros. The spokesman was aware that a jeopardy opinion had been issued, and that the pesticide had been registered. See House, *Endangered Species Act*, 1982, 408.

75 *FR* 51 (3 June 1986): 19956.

76 *FR* 43 (4 Jan. 1978): 874. On 31 Aug. 1981, Interior's solicitor issued an opinion indicating that Section 7 did not apply to federal actions in foreign countries. During the period in which overseas consultations were required, relatively few were conducted. See House, *Endangered Species Act*, 1982, 407.

77 *FR* 51 (3 June 1986): 19929. The proposed regulations, issued in 1983, had also eliminated the consultation requirement for federal actions on the high seas, but the final regulations reinstated the requirement. Ibid., 19930.

78 *Defenders of Wildlife* v. *Hodel*, 707 F.Supp. 1082 (D. Minn. 1989).

79 Science Council of Canada, *Regulating the Regulators: Science, Values and Decisions* (Ottawa: Science Council of Canada 1982), 50–51.

9 Divided Jurisdictional Responsibilities: Seeking Effective Cooperation

1 Anthony Downs, *Inside Bureaucracy* (Boston: Little, Brown, 1967), 213; Daniel A. Mazmanian and Paul A. Sabatier, *Implementation and Public Policy* (Glenview, Ill.: Scott, Foresman, 1983), 27.

2 Arthur M. Schlesinger, Jr., *The Coming of the New Deal* (Boston: Houghton Mifflin, 1959), 527–28.

3   Martin Landau, "Redundancy, Rationality, and the Problem of Duplication and Overlap," *Public Administration Review* 29 (1969): 346–58.

4   A thorough introduction to the NMFS can be found in Alfred D. Chandler, "The National Marine Fisheries Service," in William J. Chandler, ed., *Audubon Wildlife Report 1988/1989* (San Diego: Academic Press, 1988), 3–98.

5   Endangered Species Act of 1973, P.L. 93–205, Section 4(a)(2).

6   House Committee on Merchant Marine and Fisheries, hearings, *Predatory Mammals and Endangered Species*, 92nd Cong., 2nd sess., 1972, 208.

7   Letter from Maurice H. Stans, secretary of commerce, to Walter J. Hickel, secretary of the interior, 6 Nov. 1970; "Two Hickel Policies Reported Reversed," *New York Times*, 28 Nov. 1970, 25. Charles H. Meacham, commissioner for fish and wildlife, and Leslie L. Glasgow, assistant secretary of the interior for fish and wildlife and parks, were also dismissed and given three hours to vacate their offices. "Hickel Order to Protect Whales Won't be Suspended," *New York Times*, 29 Nov. 1970, 40.

8   FWS, News Release, "Green and Loggerhead Turtles Proposed for Foreign Endangered Species List," 4 Jan. 1974.

9   Lynette Stark, "Interorganizational Coordination," unpublished manuscript, 1985.

10 FWS and NMFS, "Memorandum of Understanding . . . Regarding Jurisdictional Responsibilities and Listing Procedures under the Endangered Species Act," 24 Aug. 1974. The agreement granted to the NMFS responsibility for all whales, all seals (other than walruses), and "all commercially harvested species of the phylum Mollusca and the class Crustacea which spend all of their lifetimes in estuarine waters; and all other nonmammalian species (except members of the classes Aves, Amphibia, and Reptilia), which either (i) reside the major portion of their lifetimes in marine waters; or (ii) are species which spend part of their lifetimes in estuarine waters, if the major portion of the remaining time (the time which is not spent in estuarine waters) is spend in marine waters."

11 *FR* 40 (20 May 1975): 21974.

12 FWS and NMFS, "Memorandum of Understanding Defining the Roles of the U.S. Fish and Wildlife Service and the National Marine Fisheries Service in Joint Administration of the Endangered Species Act of 1973 as to Marine Turtles," 18 July 1977; *FR* 43 (28 July 1978): 32800.

13 House Committee on Merchant Marine and Fisheries, hearings, *Fish and Wildlife Miscellaneous—Part 5*, 93rd Cong., 2nd sess., 1974, 40.

14 The NMFS receives support for protection of marine species through authorizations from both the Endangered Species Act and the Marine Mammal Protection Act of 1972. This support is combined to cover the agency's work on protected species.

15 For an excellent summary of the plight of Hawaii's birds, see J. Michael Scott, et al., "Conservation of Hawaii's Vanishing Avifauna," *BioScience* 38 (Apr. 1988): 238–53.

16 *FR* 45 (15 Dec. 1980): 82480–569.

17 *FR* 48 (21 Sept. 1983): 43104. The quoted material provides the FWS definition of high threat. Of the fifty-one Hawaiian species that the FWS ranked in 1986, thirty-two (or 62.8 percent) faced such a threat. For native species in all other

locations, 56.6 percent faced a high threat. See also Russell Stafford, "Hawaii Task Force Holds First Meeting," *Plant Conservation* 4 (Summer 1989): 1.

18   Alan D. Hart, "The Onslaught Against Hawaii's Tree Snails," *Natural History* 87 (Dec. 1978): 46. Another author has described Hawaii as "nature's grand experiment in supreme isolation." Wayne C. Gagné, "Hawaii's Tragic Dismemberment," *Defenders* 50 (Dec. 1975): 466.

19   Gagné, "Tragic Dismemberment," 466.

20   These estimates are based on the calculations of Charles Lamoureux, professor of botany, University of Hawaii.

21   Andrew J. Berger, "Hawaii's Dubious Distinction," *Defenders* 50 (Dec. 1975): 496.

22   George Cooper and Gavan Dawes, *Land and Power in Hawaii* (Honolulu: Benchmark, 1985).

23   *FR* 47 (24 Aug. 1982): 36847.

24   The material that follows is based on interviews conducted in Hawaii's DLNR in 1982.

25   Based on interviews conducted at OES, FWS, 21 Aug. 1986.

26   The Endangered Species Act Amendments of 1982, P.L. 97–304, Section 4(f).

27   OES, "Endangered and Threatened Species Recovery Priority List," computer printout provided to author on 22 Aug. 1986 and subsequently updated.

28   Some species share the same critical habitat so the total number of critical habitats is less than the number of species that benefit from these habitats.

29   *Palila v. Hawaii DLNR*, 471 F.Supp. 990 (1979).

30   Hawaii DLNR, *Hawai'i Wildlife Plan* (Honolulu, 1984), 37. The plan is an interesting document in regard to hunting; it recommends that "all public lands having significant game mammal populations . . . should be opened to controlled public hunting where feasible" (40). One possible explanation for the DLNR's responsiveness to hunters is that a large portion of its budget comes from hunters' fees.

31   *FR* 40 (16 May 1975): 21501. The critical habitat was formally designated in August 1977. See *FR* 42 (11 Aug. 1977): 40687.

32   One observer from the FWS offered a revealing observation. When he and the chief of the state's fish and game office attempted to explain to the hunters in the mid-1970s the damage the sheep were causing, "It went in one ear and out the other. . . . Some hunters were flatly opposed to any of the mountain being fenced and that all should be devoted to sheep and hunting, endangered animals and plants notwithstanding." Memorandum from Eugene Kridler, FWS/Honolulu endangered species coordinator, to director, FWS Region 1, Portland, Oregon, 26 May 1976. See also Wayne C. Gagné, "Conservation Priorities in Hawaiian Natural Systems," *BioScience* 38 (Apr. 1988): 268.

33   National Wildlife Federation, *Conservation Report*, 3 Oct. 1975, 438.

34   Letter from Lynn Greenwalt, director, FWS, to Michael R. Sherwood, Sierra Club Legal Defense Fund, 18 May 1978.

35   *Palila v. Hawaii DLNR* (1979).

36   *Palila v. Hawaii DLNR*, 639 F.2d 495 (1981).

37   *Palila v. Hawaii DLNR*, 649 F.Supp. 1070 (D. Hawaii 1986). The state appealed the decision to the U.S. Court of Appeals, which affirmed the lower court's verdict in favor of the palila. See *Palila v. Hawaii DLNR*, 852 F.2d 1106 (9th Cir. 1988).

38 An excellent summary of the political status of the Hawaiian monk seals can be found in Michael R. Sherwood, staff attorney, Sierra Club Legal Defense Fund, "Memorandum of Points and Authorities in Support of Motion for Summary Judgment" (1987). The plaintiffs filed this memorandum in *Greenpeace International, Inc. et al.* v. *Baldridge* in the U.S. District Court of Hawaii.

39 Karl W. Kenyon, "No Man is Benign: The Endangered Monk Seal," *Oceans* 13 (May-June 1980): 48.

40 Ibid., and letter from John R. Twiss, Jr., executive director, MMC, to Robert W. Schoning, director, NMFS, 24 Dec. 1975.

41 Twiss letter, 24 Dec. 1975.

42 Marine Mammal Protection Act of 1972, 16 U.S.C., Sections 1401–1407.

43 FWS, *Hawaiian Islands National Wildlife Refuge: Master Plan/Environmental Impact Statement* (Washington, 1986), 425.

44 Hawaii DLNR, *Hawaii Fisheries Development Plan* (Honolulu, 1979).

45 *FR* 41 (23 Nov. 1976): 51611. The seal had earlier been designated as a depleted species under the provisions of the Marine Mammal Protection Act of 1972. See *FR* 41 (22 July 1976): 30120. The state's position on the desirability of listing the monk seal is somewhat ambiguous. The governor endorsed the proposal to list the seal in 1976, but the director of the DLNR Division of Fish and Game later said that the state was "apprehensive about the listing, which we believed was based more on emotionalism than on the actual status of the Hawaiian seal population. . . ." Letter from Kenji Ego, director, Division of Fish and Game, Hawaii DLNR, to Henry Eschwege, GAO, 4 Mar. 1981.

46 NMFS, Southwest Region, "Hawaiian Monk Seal Management Plan," (Sept. 1976).

47 *FR* 41 (23 Nov. 1976): 51611.

48 Letter from John R. Twiss, Jr., executive director, MMC, to Terry L. Leitzel, assistant administrator for fisheries, NMFS, 6 Apr. 1979.

49 NMFS, *Draft Environmental Impact Statement for the Proposed Designation of Critical Habitat for the Hawaiian Monk Seal*, (Washington, 1980), 55–56.

50 Memorandum from Doyle Gates, Pacific Program office administrator, Southwest Region, NMFS, to E. C. Fullerton, director, Southwest Region, NMFS, 30 Mar. 1984.

51 Letter from Susumu Ono, chair, Hawaii Board of Land and Natural Resources, to Richard B. Roe, director, Office of Protected Species and Habitat Conservation, NMFS, 19 Feb. 1985.

52 NMFS, *Final Environmental Impact Statement: Proposed Designation of Critical Habitat for the Hawaiian Monk Seal in the Northwestern Hawaiian Islands* (Washington, 1986).

53 *FR* 51 (30 Apr. 1986): 16047.

54 Letter from Michael R. Sherwood, staff attorney, Sierra Club Legal Defense Fund, to the author, 19 Jan. 1988.

55 *FR* 53 (26 May 1988): 18988–90. Not surprisingly, the state of Hawaii objected to the change, claiming that the best available scientific evidence did not justify the designation of a critical habitat.

56 John W. Kingdon, *Agendas, Alternatives, and Public Policies* (Boston: Little, Brown, 1984), 115.

57 State of Hawaii, Endangered Species Act of 1975, Hawaii Revised Statutes, Section 195D–1.

58   Craig S. Harrison, "A Marine Sanctuary in the Northwestern Hawaiian Islands: An Idea Whose Time Has Come," *Natural Resources Journal* 25 (Apr. 1985): 329.

59   Letter from John P. Craven, marine affairs coordinator, Office of the Governor, State of Hawaii, to John R. Twiss, Jr., executive director, MMC, 12 Apr. 1978.

60   Memorandum from E. C. Fullerton, southwest regional director, NMFS, to William G. Gordon, assistant administrator for fisheries, NMFS, 15 May 1984. For an indication of the state's opposition to designation of critical habitat, see letter from Kenji Ego, director, Division of Fish and Game, Hawaii DLNR, to Gary Smith, chief, Fisheries Management Division, Southwest Regional Office, NMFS, 27 Nov. 1978.

61   Memorandum from David Cottingham, Office of the Administrator, National Oceanic and Atmospheric Administration, to Bob Brumsted, chief, Protected Species Divisions, NMFS, 28 Aug. 1985.

62   The FWS had known for several months before the president's signing of the Endangered Species Act of 1973 that Commerce would have a role in listing some marine species. If the FWS was concerned about Commerce's preferences, the Service could easily have proposed and listed the turtles before the law was signed.

63   *Palila v. Hawaii DLNR* (1979), 989.

64   *FR* 46 (2 June 1981): 29490.

65   *FR* 46 (4 Nov. 1981): 54748. As one FWS employee later suggested during an interview with the author, efforts to change the definition of "take" represented "a highly political attempt to rewrite the Endangered Species Act on the part of political appointees" in Interior's Office of the Solicitor.

10   Returning Endangered Species to Better Times

1   Rollin D. Sparrow and Howard M. Wight, "Setting Priorities for the Endangered Species Program," in *Transactions of the 40th North American Wildlife and Natural Resources Conference* (Pittsburgh: Wildlife Management Institute, 1975), 153.

2   David Marshall, "Endangered Species Priority System," 11. Marshall's paper is an internal document of the FWS, written in 1976. The FWS never adopted Marshall's proposed priority scheme.

3   FWS, *Endangered and Threatened Species Recovery Planning Guidelines* (Washington, n.d.), 3.

4   House Committee on Merchant Marine and Fisheries, hearings, *Endangered Species Oversight*, 94th Cong., 1st sess., 1975, 47–48; GAO, *Endangered Species—A Controversial Issue Needing Resolution* (Washington, 1979), 68.

5   Other species with long delays include the red-cockaded woodpecker, the Puerto Rican parrot, the Laysan duck, and the light-footed clapper rail. The dates for FWS receipt of the technical drafts for these species can be found in House Committee on Merchant Marine and Fisheries, hearings, *Endangered Species—Part 1*, 95th Cong., 2nd sess., 1978, 128–33. The dates for final approval of the plans can be found in *ESTB* 10 (Jan. 1985): 3–6.

6   Endangered Species Act of 1973, as amended, Sections 4(g) and 4(g)(4).

7   Ibid., Section 4(f)(1).

8   House Committee on Merchant Marine and Fisheries, *Endangered Species Act*

Amendments—Part 1, 97th Cong., 2nd sess., 1982, H. Rept. 97–567, 55. The Department of the Interior shared these concerns. See ibid., 41.

9   For discussion of the proposed system, see *FR* 48 (19 Apr. 1983): 16756–59. The final notice is in *FR* 48 (21 Sept. 1983): 43098–105.

10 FWS, *Endangered Species Program Management Document* (Washington, 1980), 28.

11 Memorandum, "Comments on Endangered and Threatened Species Listing and Recovery Guidelines," from William F. Shake, assistant regional director, Federal Assistance, FWS Region 1, Portland, Oregon, to Ronald Lambertson, associate director, Federal Assistance, FWS, Washington, 26 May 1983. Shake's counterpart in FWS Region 6 (Denver) shared identical concerns.

12 *FR* 48 (21 Sept. 1983): 43104.

13 Letter from Michael J. Bean and Bruce S. Manheim, Environmental Defense Fund, to Ronald Lambertson, associate director, Federal Assistance, FWS, 16 June 1983.

14 *ESTB* 1 (Aug. 1976): 2; and Shake, "Comments on Endangered and Threatened Species Listing."

15 House, *Endangered Species—Part 1*, 8.

16 House Committee on Merchant Marine and Fisheries, *Endangered Species Act Amendments of 1988*, 100th Cong., 2nd sess., 1988, H. Rept. 100–928, 2–3.

17 Sierra Club Legal Defense Fund, *In Brief* (Spring 1987), 4. In explaining its lack of progress in developing recovery plans, the NMFS said its activity reflects the low priority it assigns to endangered species. See GAO, *Endangered Species: Management Improvements Could Enhance Recovery* (Washington, 1988), 25, 89.

18 *ESTB* 10 (May 1985): 7. The woodpecker may still exist in Cuba, but only in negligible numbers.

19 These include the Arizona hedgehog cactus (listed 25 Oct. 1979), Lloyd's hedgehog cactus (listed 26 Oct. 1979), Lloyd's mariposa cactus, the bunched cory cactus (both listed 6 Nov. 1979), and the malheur wire lettuce (listed 11 Nov. 1982). All these species are plants. The sixth species is the Scioto madtom, a fish listed on 25 Sept. 1979. The latter species was presumed to be extinct in 1988. Paradoxically, its recovery potential in 1986 was classified as high.

20 These include the reed warbler (listed 2 June 1970), which is now presumed to be extinct, and four Hawaiian plants: *lipochaeta venosa, stenogyne angustifolia, haplostachys haplostachya* (all listed 30 Oct. 1979), and the 'Ewa plains 'akoko (listed 24 Aug. 1982).

21 House Committee on Merchant Marine and Fisheries, hearings, *Fish and Wildlife Miscellaneous—Part 3*, 98th Cong., 1st sess., 1983, 61–62.

22 Senate Committee on Environment and Public Works, *Endangered Species Act Amendments of 1987*, 100th Cong., 1st sess., 1987, S. Rept. 100–240, 4.

23 House Committee on Merchant Marine and Fisheries, hearings, *Endangered Species Act*, 99th Cong., 1st sess., 1985, 43.

24 FWS, *Endangered Species Program Management Document* (Washington, 1980), 29. Analyzing the FWS's annual report on the implementation of recovery actions, Defenders of Wildlife estimated that 84 of 165 recovery plans were being implemented in 1984. Defenders of Wildlife, *Saving Endangered Species: Amending and Implementing the Endangered Species Act* (Washington: Defenders of Wildlife, 1986), 16. In response to such data, the Endangered Species Act Amendments

of 1988 require the FWS and NMFS to provide biennial reports to Congress "on the status of efforts to develop and implement recovery plans for all species listed . . . and on the status of all species for which such plans have been developed."

25 *FR* 48 (21 Sept. 1983): 43104.
26 Senate Committee on Appropriations, hearings, *Department of the Interior and Related Agencies Appropriations for Fiscal Year 1986*, 99th Cong., 1st sess., 1985, 821.
27 Memorandum, "Analysis of Future Recovery Costs—Table III for the Santa Cruz long-toed salamander and blunt-nosed leopard lizard," from Richard J. Navarre, acting project leader, Sacramento Endangered Species Office, FWS, to director, OES, 12 July 1984.
28 FWS, OES, "Threshold Species: Included on [Program Manager's] FY87 PMSS," 1986. See also Senate, *Department of the Interior . . . Appropriations for Fiscal Year 1986*, 822.
29 S. Rept. 100–240, *Endangered Species Act Amendments of 1987*, 9.
30 Ibid.
31 Marshall, "Endangered Species Priority System."
32 GAO, *Endangered Species: Management Improvements*, 33. This extra attention seems to have been rewarded. In February 1990 the FWS announced a status review of the bald eagle in anticipation of a proposal to reclassify or delist the species. See *FR* 55 (7 Feb. 1990): 4209.
33 The episodes involving the woodpeckers and cactus are described in Peggy Olwell, "The Knowlton Cactus," in Roger L. Di Silvestro, ed., *Audubon Wildlife Report 1986* (New York: National Audubon Society, 1986), 946; and Di Silvestro, "Endangered Species? Load the Shotgun!" *Audubon* 89 (Sept. 1987): 12. The isopods and ladies' tresses are discussed in GAO, *Endangered Species: Management Improvements*, 76–78, 84–86.
34 GAO, *Endangered Species—A Controversial Issue*, 73, 75. There is considerable irony in this statement inasmuch as the GAO criticized the FWS in 1988 for incorporating such nonbiological variables into the recovery process. See GAO, *Endangered Species: Management Improvements*, ch. 4.
35 House, *Endangered Species Oversight*, 257. For a discussion of the early history of the grant program, see *ESTB* 4 (Dec. 1979): 3–5; and *ESTB* 6 (Oct. 1981): 6–7.
36 Defenders of Wildlife, *Saving Endangered Species: Implementation of the Endangered Species Act, 1987* (Washington: Defenders of Wildlife, 1987), 13.
37 *ESTB* 6 (Oct. 1981): 7. These percentages summarize expenditures through fiscal year 1981. In addition to projects related to federally listed species, Section 6 funds are also available for state-listed or candidate species. No funds were available for state grants in fiscal year 1982.
38 Hank Fischer, "Deep Freeze for Wolf Recovery?" *Defenders* 62 (Nov.-Dec. 1987): 30. For a well-written history of public policies toward wolves in America, see Thomas R. Dunlap, *Saving America's Wildlife* (Princeton, N.J.: Princeton Univ. Press, 1988).
39 Fischer, "Deep Freeze," and Fischer, "Wolves for Yellowstone," *Defenders* 63 (Mar.-Apr. 1988): 16.
40 An Interior employee provided details of the meeting to the author.
41 FWS, *Palos Verdes Blue Butterfly Recovery Plan* (1984), 7, 9.
42 Ibid., 9.

43   Ibid., 21.

44   Michael Wines, "Official Bungling May Have Erased Rare Species," *Los Angeles Times*, 11 June 1984, sec. 1, 3.

45   Ibid. In response to the city's alleged contribution to the "taking" and extinction of the butterflies, the U.S. government filed a criminal action against Rancho Palos Verdes. The city's defense, which was convincing to the courts, was that the Endangered Species Act's prohibition on taking applied only to persons, not cities. See *United States* v. *City of Rancho Palos Verdes* 841 F.2d 329 (9th Cir. 1988). The Endangered Species Act Amendments of 1988 revised the definition of person to indicate that cities would henceforth be subject to the law's provisions.

46   FWS, *California Condor Recovery Plan* (1974), 10.

47   As examples, see Gar Smith, "Science Fails the Condor," *San Francisco Chronicle*, 23 Feb. 1986; and Mark J. Palmer, "Too Late for the Condor?" *Sierra* 71 (Jan.-Feb. 1986): 33–38. All this attention should not suggest that the recovery effort proceeded without conflict. Throughout much of the effort the FWS, the California Fish and Game Commission, and the National Audubon Society disagreed about what should be done. The Audubon Society unsuccessfully sued the FWS in an effort to prevent removal of the last condors from the wild.

48   In early 1990, thirty condors remained, about equally divided between the Los Angeles Zoo and the San Diego Wild Animal Park. This total includes a few chicks that were born in early 1989. The goal is to release some of the condors in the 1990s.

49   *FR* 50 (12 Sept. 1985): 37192–94; GAO, *Endangered Species: Management Improvements*, 18. The birds are the Palau dove, the Palau owl, and the Palau fantail. The population of the brown pelican found in Alabama, Florida, and on the Atlantic coast has also been delisted due to its recovery, most likely because of the nationwide ban on DDT. The pelican is still listed as endangered throughout the rest of its range, which includes California, Texas, Louisiana, Mississippi, and the coastal areas of Central and South America and the West Indies. On the alligator, see George Laycock, "The Unendangered Alligator's Rapid Fall from Grace," *Audubon* 89 (Sept. 1987): 39–43. The alligator remains listed as a threatened species because of the similarity of its appearance to other listed species. The species whose status has been changed from endangered to threatened include the apache trout (status changed in 1975, recovery plan approved in 1979); the paiute cutthroat trout (1975, plan approved in 1985); the greenback cutthroat trout (1978, plan approved in 1979); the snail darter (1984, plan approved in 1983); the arctic peregrine falcon (1984, no recovery plan prior to change in status); the Utah prairie dog (1984, no recovery plan prior to change in status); the Tinian monarch flycatcher (1987, no recovery plan prior to change in status); and the Aleutian Canada goose (1990, plan approved in 1979). The populations include the gray wolves in Minnesota and bald eagles in Oregon, Washington, Minnesota, Michigan, and Wisconsin.

50   These species include: the Tecopa pupfish (delisted in 1982); the blue pike, the longjaw cisco, and the Santa Barbara song sparrow (all delisted in 1983); Sampson's pearly mussel (delisted in 1984); and the amistad gambusia, a fish (delisted in 1987).

51   Species whose continued existence is in doubt include the Mariana mallard, the

Mariana fruit bat, the Palos Verdes blue butterfly, the giant anole, the tubercled-blossom mussel, the turgid-blossom mussel, the yellow-blossom mussel, the eastern cougar, the dusky seaside sparrow, Bachman's warbler, the White Cat's paw pearly mussel, the Scioto madtom, the Guam broadbill, the reed warbler, the Molokai creeper, the bridled white eye, and the Micronesian kingfisher. Through early 1990, none of these species had been proposed for delisting. See GAO, *Endangered Species· Management Improvements*, 21; and CEQ, *Environmental Quality, 1986* (Washington, 1988), 34–35.

52 *FR* 49 (9 Nov. 1984): 44775.
53 *ESTB* 8 (July 1983): 5.

11 Assessing Performance and Its Consequences

1 This is the author's impression after extensive discussion and correspondence with many FWS employees over a period of several years. See also Philip Shabe-coff, "At the Wildlife Service: An Extinction with Irony," *New York Times*, 25 Nov. 1987, B8.

2 See Senate Committee on Environment and Public Works, *Endangered Species Act Amendments of 1987*, 100th Cong., 1st sess., 1987, S. Rept. 100–240, 4.

3 Endangered Species Act of 1973, Appropriation Authorization, P.L. 96–159 (1979), Section 3(6).

4 S. Rept. 100–240, *Endangered Species Act Amendments of 1987*, 4.

5 FWS and NMFS, "Critical Habitat Priorities," mimeographed, 17 Feb. 1978.

6 Anthony Downs, "Up and Down with Ecology—The 'Issue-Attention Cycle,'" *The Public Interest* 28 (Summer 1972): 38–50; and John W. Kingdon, *Agendas, Alternatives, and Public Policies* (Boston: Little, Brown, 1984), 108–10.

7 Ibid., 110.

8 Harold Sprout and Margaret Sprout, *The Context of Environmental Politics* (Lexington: Univ. of Kentucky Press, 1978), 168.

9 David A. Stockman, *The Triumph of Politics* (New York: Harper & Row, 1986), 376.

10 S. Rept. 100–240, *Endangered Species Act Amendments of 1987*, 13.

11 Defenders of Wildlife, *Saving Endangered Species* (Washington: Defenders of Wildlife, 1987), 16.

12 There have been some efforts to move in this direction. In mid-1988 the National Biological Diversity Conservation and Environmental Conservation Act (H.R. 4335) was introduced into the House. The proposal, which was not approved, would have created a National Biodiversity Research Center and amended the NEPA to require consideration of the effects of federal actions on biological diversity. See House Committee on Science, Space, and Technology, hearings, *National Biological Diversity Conservation and Environmental Conservation Act*, 100th Cong., 2nd sess., 1988.

13 Paul Ehrlich and Anne Ehrlich, *Extinction: The Causes and Consequences of the Disappearance of Species* (New York: Random House, 1981), 242.

14 Lettie McS. Wenner, *One Environment Under Law* (Pacific Palisades, Calif.: Goodyear, 1976), 123–29.

15 See, for example, Ehrlich and Ehrlich, *Extinction*, ch. 10; William Ophuls, *Ecology*

and the Politics of Scarcity (San Francisco: W. H. Freeman, 1977); and Robert L. Heilbroner, *An Inquiry into the Human Prospect* (New York: W. W. Norton, 1974).

16   Heilbroner, *An Inquiry*, 38.

17   Norman Myers, "The Exhausted Earth," *Foreign Policy* 42 (Spring 1981): 53. A related discussion is found in World Commission on Environment and Development, *Our Common Future* (New York: Oxford Univ. Press, 1987), 176.

18   Gaylord Nelson, "Ah, Wilderness! Save It," *New York Times* 4 Sept. 1984, A21.

Index

About the Author

Richard J. Tobin is Associate Professor of Political Science, State University of New York at Buffalo. He is the author of *The Social Gamble: Determining Acceptable Levels of Air Quality* and has published numerous articles in *Environmental Affairs*, *Environmental Conservation*, the *American Journal of Political Science*, and other journals.

Library of Congress Cataloging-in-Publication Data
Tobin, Richard J.
The expendable future : politics and the protection of biological diversity / Richard J. Tobin.
ISBN 0-8223-1053-8. — ISBN 0-8223-1071-6 (pbk.)
1. Biological diversity conservation—Environmental aspects—United States.   2. Biological diversity conservation—Political aspects—United States.   I. Title.
QH76.T63   1990
333.95—dc20   90-2867